Campden & Chorleywood Food
n Group

Key Topics in Food
Science and Technology – No. 6

Food chemical composition: dietary significance in food manufacturing

Tim Hutton

Campden & Chorleywood Food Research Association Group comprises
Campden & Chorleywood Food Research Association
and its subsidiary companies
CCFRA Technology Ltd   CCFRA Group Services Ltd   Campden & Chorleywood Magyarország

© CCFRA 2002

Campden & Chorleywood Food
Research Association Group

Chipping Campden, Gloucestershire, GL55 6LD UK
Tel: +44 (0) 1386 842000   Fax: +44 (0) 1386 842100
www.campden.co.uk

Published by CCFRA in collaboration with the Royal Society of Chemistry

Information emanating from this company is given after the exercise of all reasonable care
and skill in its compilation, preparation and issue, but is provided without liability in its
application and use.

Legislation changes frequently. It is essential to confirm that legislation cited in this
publication and current at the time of printing is still in force before acting upon it. Any
mention of specific products, companies or trademarks is for illustrative purposes only and
does not imply endorsement by CCFRA or RSC.

ROYAL SOCIETY OF CHEMISTRY

Thomas Graham House, Science Park, Milton Road, Cambridge, CB4 0WF, UK.
Tel: +44 (0)1223 432360 Fax: +44 (0)1223 426017 www.rsc.org and www.chemsoc.org
Registered Charity No. 207890

© CCFRA 2002
ISBN: 0 905942 50 7
A catalogue record for this book is available from the British Library.

SERIES PREFACE

Food and food production have never had a higher profile, with food-related issues featuring in newspapers or on TV and radio almost every day. At the same time, educational opportunities related to food have never been greater. Food technology is taught in schools, as a subject in its own right, and there is a variety of food-related courses in colleges and universities - from food science and technology through nutrition and dietetics to catering and hospitality management.

Despite this attention, there is widespread misunderstanding of food - about what it is, about where it comes from, about how it is produced, and about its role in our lives. One reason for this, perhaps, is that the consumer has become distanced from the food production system as it has become much more sophisticated in response to the developing market for choice and convenience. Whilst other initiatives are addressing the issue of consumer awareness, feedback from the food industry itself and from the educational sector has highlighted the need for short focused overviews of specific aspects of food science and technology with an emphasis on industrial relevance.

The *Key Topics in Food Science and Technology* series of short books therefore sets out to describe some fundamentals of food and food production and, in addressing a specific topic, each issue emphasises the principles and illustrates their application through industrial examples. Although aimed primarily at food industry recruits and trainees, the series will also be of interest to those interested in a career in the food industry, food science and technology students, food technology teachers, trainee enforcement officers and, established personnel within industry seeking a broad overview of particular topics.

Leighton Jones
Series Editor

PREFACE TO THIS VOLUME

Food, like the many living things from which it is derived, is made of chemicals. To the human body, the vast majority of these are beneficial or innocuous, though some present a hazard. To the food manufacturer, however, all the chemicals present in a food or raw material are potentially important: they determine its nutritional value, eating properties and suitability for use in particular products and processes. With this in mind, this short book sets out to explain to those without expertise in food chemistry, some of the basics of food chemical composition. But with several good books on food composition already available, why produce another? There are several good reasons.

First, this book has a strong industrial slant: it uses examples from food manufacturing and the industry-consumer interface to put food composition in context. It then relates aspects of composition to wider issues like safety assurance, traceability, product development and labelling. Second, it emphasises that all that we eat is made from chemicals: many 'good', some 'bad' and most 'indifferent'. Finally, this is a short book whereas industrial food chemistry is an enormous subject - so this book is not comprehensive. It uses selected examples to illustrate points that often get overlooked in discussions of the chemicals natural to foods or used in preservation and processing.

As the primary role of our food is to provide the human body with nutrients and energy, the book loosely groups the food components covered in terms of dietary significance: 'nutrients', 'non-nutrients', and those implicated in special dietary needs. However, it is a book about composition not nutrition, and so does not attempt to deal with food-related health issues such as obesity and high blood pressure.

Tim Hutton

ACKNOWLEDGEMENTS

Thanks are due to Drs. Helen Brown, Anton Alldrick and Leighton Jones for their valuable comments on the scientific accuracy of the text.

NOTE

All definitions, legislation, codes of practice and guidelines mentioned in this publication are included for the purposes of illustration only and relate to UK practice unless otherwise stated.

CONTENTS

1. INTRODUCTION

1.1 What is food made of?

The simple answer to this is chemicals – and lots of them. Food is derived from plants, animals and microbes which are, in essence, highly organised chemical systems. Many of the chemicals within food are essential for human life – others just happen to be associated with the material we actually want to eat. Some (flavour and texture components and colour) actually persuade us to eat the food. We choose to eat certain foods because of a combination of the chemicals they contain that give it a pleasant taste and appearance and those that we actually need in order to survive

In many instances, especially with animal-based products, many of the major chemicals that are present in the food serve the same or similar purpose in human health and maintenance of life as they did in the 'food' when it was living. This is an important concept to bear in mind when looking at the individual components of foods and our own nutritional requirements.

The types of constituents of foods can be broadly summarised as follows:

- carbohydrates, fats and proteins - the macronutrients - to provide energy and the building blocks for growth, development and survival;

- vitamins and minerals which are needed in small amounts for the body to be able to function properly;

- fibre (along with other constituents) to assist in digestion and other functions;

- flavour, texture and colour compounds, which generally serve no nutritional purpose, but make the food appear and taste good;

- other compounds that either have no function at all as far as we are concerned, or are deleterious to us, but were acquired by the plant or animal during its lifetime and are unavoidably associated with the food; and

- other chemicals (e.g. preservatives, stabilisers, emulsifiers), added to achieve the desired properties of the food

We have developed knowledge to eat those materials that best provide the chemicals we need, but all foods will naturally contain substances that we do not need, and in some cases some that we would rather not consume at all. It is also important to consider that many chemicals may be of no consequence in isolation, but may become important (either in a good or bad way) when in combination with others.

1.2 The role of production and processing operations

There has been much talk in recent years about the nutritional aspects of processed versus unprocessed food, and primary food (e.g. meat, fish, fruit, vegetables) versus formulated food (e.g. bread and cakes, cheese, and ready meals). A detailed discussion of the nutritional qualities of various food groups is outside the scope of this book. However, it is intended to demonstrate the chemical similarities between many of these food types, and that fresh or unprocessed foods do not always have a magical set of ingredients that are totally lacking in formulated and processed foods. Many of the flavourings and 'artificial' additives that are incorporated into some formulated food may be found naturally in raw or unprocessed food products.

It is true that formulation and processing will change the chemical make-up of food, often quite subtly, and that some of these changes will be generally beneficial while others are unfavourable. The food manufacturer makes great efforts to maximise the beneficial effects and minimise the unfavourable ones.

Cooking and different processing techniques can significantly alter the chemical composition of food, as well as its physical and sensory properties. In any process, it is likely that some benefits will be traded off against some deleterious changes. Changes will also occur during storage of unprocessed food - again, some of these changes will be desirable, whereas others will not, and the key is to minimise the undesirable changes (e.g. sugar conversion to starch in peas), while taking advantage of the desirable changes (e.g. ripening of some fruits, and the effects of 'hanging' of meat post mortem).

The composition of the raw material may also show significant variation. With fruit and vegetables, for example, not only will the relative level of individual components vary with variety type, but environmental conditions may exert a significant effect. Factors such as soil type and level of added fertilizer, amount of rainfall, timing of rainfall, amount of sunshine, time of harvest (i.e. maturity), and length of storage period will all influence food chemical composition. Comparable differences may be found with foods of animal origin. For example, beef and veal, although derived from the same type of animal, are very different in both appearance and flavour.

1.3 The significance of food composition

The nature and occurrence of the different chemical components of food, and the steps that food manufacturers and processors have to take in relation to the dietary aspects of food is the focus of this Key Topic. It begins with an overview of the chemistry of the major groups of nutrients and their role in the body, and follows this with a similar discussion of non-nutrients such as flavour components, antinutrients and toxicants found in food. Using this as a base, it then explores some common and well established dietary related conditions and describes how food manufacturers, processors and retailers approach these, and their implications for product labelling, as part of the industrial activity of food production.

The role of food chemicals in industrial food production is largely outside the scope of this book, except where it impinges upon and is itself affected by dietary aspects. An in-depth discussion of nutrition itself and related matters such as obesity is also outside the scope of the book. There are many textbooks that deal with nutrition, and Whitney *et al* (1998) is recommended further reading.

2. NUTRIENTS – THEIR CHEMICAL NATURE AND ROLE IN THE BODY

The primary purpose of eating food is to obtain the nutrients that it provides. Historically, processing has been used to make food more suitable for consumption in one way or another. As well as destroying some undesirable chemicals in food, processes such as cooking can also make nutrients more 'available' for digestion, by changing their structure or by releasing them from other components of the food. Thus, although processing can result in loss of nutrients (especially some vitamins and minerals), it can also make food more nutritious. Processing can also make food more desirable, by altering its flavour, texture and colour, and can enable it to be stored for longer: fermentation, pickling, drying and freezing are all traditional 'shelf-life extension' techniques. The end result is greater availability in the long term of nutrients.

Most of the nutrients in food fall into two main categories. The macronutrients - fats, proteins and carbohydrates - provide the fuel and building blocks for production and maintenance of body mass and life. The micronutrients - vitamins and minerals - are required in small amounts by the body for many functions, where their specific chemical and molecular properties are utilised. In addition, fibre and water can be regarded as nutritional components: fibre facilitates normal gut function and has some significant effects on body metabolism, and water is required by all living cells for virtually all metabolic activities.

Table 1 illustrates the levels of the macronutrients and fibre, and total energy value of some typical, but quite different foods. In addition, the types and levels of vitamins and minerals will vary from one product to another, and some of the significant components are also listed (Table 2). The contribution that all of these components make to the diet, especially the micronutrients, will depend on the volume of the food eaten, the availability of the components (i.e. how easily they are digested and absorbed), and in some cases how they respond to various processing regimes.

Table 1 - Major nutritional ingredients of selected foods (per 100g)
(Holland *et al.*, 1991)

Food type	Water g	Prot g	Fat g	Carb g	Fibre(*) g	Energy kcal
Rump steak, raw	66.7	18.9	13.5	0	0	197
White bread	37.3	8.4	1.9	49.3	1.5	235
New potatoes, raw	81.7	1.7	0.3	16.1	1.0	70
Cheddar cheese	36.0	25.5	34.4	0.1	0	412
Eggs, raw	75.1	12.5	10.8	Trace	0	147
Apples, raw	84.5	0.4	0.1	11.8	1.8	47

(* by the Englyst method)

Table 2 - Major vitamins and minerals
(Holland *et al.*, 1991)

Rump steak, raw	Potassium, Niacin
White bread	Calcium, Phosphorus, Niacin
New potatoes, raw	Potassium, Vitamin C, Vitamin B6
Cheddar cheese	Calcium, Phosphorus, Sulphur, Vitamins A, D, E and B12, Riboflavin
Eggs, raw	Vitamins A, B12, D and E, Riboflavin, Folate, Pantothenate, Biotin
Apples, eating, raw	Vitamin C

There is no such thing as a healthy or unhealthy food (assuming that it does not contain significant toxin levels) - it is combinations of different foods that might give rise to a healthy or unhealthy diet. Given the trend towards more convenience foods, there has been much consumer pressure in recent years for formulated foods to be even more nutritious to facilitate a healthy diet. In some cases this can be achieved by using novel combinations of processes that involve less heat processing, in order to conserve those nutrients that are unstable to heat. The food industry has put considerable effort into matching consumer requirements, reformulating products to be higher in fibre or protein, enriched in vitamins and/or minerals, and lower in fat and total energy (principally fat and carbohydrate). This is reflected in the number of claims made on the labels of food packaging. For the consumer who is conscious about their health and the food that they eat, there is a wide choice of prepared foods to suit their requirements. For example, eating a high fibre diet has long been recommended as having a beneficial effect on health and many products now have added fibre or are marketed as being high in fibre. The 'healthy' products have the appropriate health statement boldly printed on the front of the packaging. For example:

Added Vitamins & Minerals
Low Calorie
Low Fat
Low Salt/Sodium
Low Sugar

In some cases, the product may always have had the charactersitic being highlighed, e.g. low in fat. However, manufacturers now actively promote these facts, and some also identify these ranges with distinctive sub brand names (see box 1).

In addition to the formulated foods developed with 'healthy eating' in mind, there has been a significant increase in the variety of 'minimally processed' foods available. These are typically chopped fruits, vegetables and salad mixes, often packaged under a modified atmosphere (i.e. a combination of oxygen, carbon dioxide and nitrogen different to that found in air) and stored chilled, to prolong the shelf life. These combine the 'healthy' aspect of raw fruit and vegetable products, with extra convenience and the capacity for longer storage.

Box 1 - Some 'healthy eating' sub-brands (2001)

Own Label
Be Good to Yourself – Sainsbury's
Healthy Selection – Somerfield/Budgens
Healthy Living CWS
Healthy Eating – Tesco
Count on us… - Marks & Spencer
Healthy Choice – Asda/Safeway

Branded
Shapers – Boots Shape – St Ivel Ltd
Müller Light – Müller Dairy
Go Ahead – United Biscuits
Healthy Menu – Rye Valley Foods
So Good – So Good International Ltd
Weight Watchers from Heinz - Heinz and other manufacturers under licence

(Llewellyn-Davies, 2001).

It is worth emphasising again, however, that many raw foods are inedible in the raw state and need to be processed in some way. Staples such as potatoes contain starch that is not digestible in the raw state, whilst many legumes contain similarly indigestible proteins, and others contain chemicals that are toxic and need to be destroyed by processing. Producing food with the maximum amount of nutrients remaining, but with unwanted components removed, has been learnt empirically from the beginning of mankind. However, our knowledge and understanding of different ways to achieve this continues to expand.

2.1 Carbohydrates

Carbohydrates are the principal source of energy in the human diet. Although they do not provide as much energy on a weight for weight basis as fats, the volume of carbohydrate we take in compensates for this. Carbohydrates come in many forms, from the simple sugars to complex carbohydrates (polysaccharides), which are made up of many linked sugar units. In simple terms we can utilise the sugars and some complex carbohydrates (primarily starch), but not others (e.g. cellulose).

Box 2 - Energy values of the macronutrients

The different macronutrients in food have different energy values. These are used by food manufacturers to calculate the total energy content of the food. For consistency of labelling, the figures to be used in these calculations are laid down in law in the EU. Figures have to be quoted both in kilojoules (kJ) and kilocalories (kcal).

Component	Energy value (per gram)	
	kJ	kcal
Fat	37	9
Carbohydrate	17	4
Protein	17	4
Polyols	10	2.4
Ethanol	29	7
Organic acids	13	3

The building blocks of carbohydrates are the simple sugars - monosaccharides - which are made up of carbon, hydrogen and oxygen. There are three major monosaccharides, glucose, fructose and galactose, which are important in nutrition. For energy-generating purposes they can all be considered to be equivalent to glucose, and are metabolised, ultimately, to carbon dioxide and water. There are various mechanisms for this to happen. The detailed biochemistry behind these is beyond the scope of this publication, but there are many textbooks that deal with it (e.g. Stryker, 1988).

A fairly constant supply of carbohydrate in the diet is important; the brain, other nerve cells and developing red blood cells require glucose as an energy source and cannot function on fats as a sole energy source. Without dietary carbohydrate, protein would have to be converted to glucose to fulfil this need (Whitney *et al*, 1998)

Box 3 - Molecular structure of glucose

The structure of monosaccharides can be depicted in several ways, but it should be remembered that in reality they are 3-dimensional. In this depiction of glucose ($C_6H_{12}O_6$), the standard numbering system for the carbon atoms is shown (see later for explanation on the structure of polysaccharides in general and starch in particular).

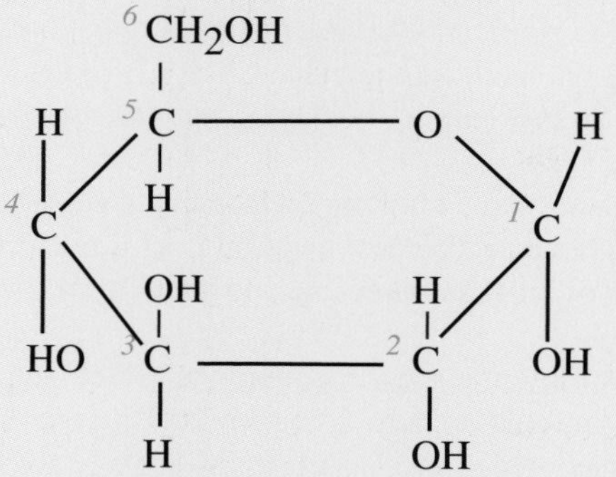

$$
\begin{array}{c}
{}^6CH_2OH \\
|
\end{array}
$$

H ^{5}C ——————— O ╲   H

^{4}C ^{1}C

HO ^{3}C ——————— ^{2}C   OH

H   OH

We take in carbohydrates in many forms; the primary forms are starch, lactose, sucrose and glucose and other monosaccharides. The overwhelming proportion of our carbohydrate intake comes directly from plant sources. In bulk terms, we take in most in the form of starch, although many fruits have high sugar contents and little or no starch.

Starch is a polysaccharide, made up of many glucose units joined together in complex chains, some of which are branched. Starch is the major form in which most plants store carbohydrates. Being such a large molecule, there is much scope for variations in structure and different plants are characterised by different starch molecular and granular structures. In fact starch is not a 'single' molecule, but is made up of two distinct components: amylose and amylopectin. Amylose is a long, virtually unbranched chain of glucose molecules linked through the carbon atoms at positions 1 and 4. Amylopectin is a highly branched polymer with 15-30 of the (1-4) linkages in each branch, the branches being joined by linkages between the carbon atoms at positions 1 and 6.

Box 4 - Reduced sugar products

There has been much pressure in recent years for a reduction in the level of sugars added to foods, both because of general dietary concerns and because of the role that sugars play in dental caries. The banner 'no added sugar' is now often seen on product labels. In 1999 and 2000, 88 new products with this banner (mainly drinks, but also breakfast cereals, bakery products and canned products) were identified by CCFRA (NewFoods, 2001).

In some cases, reducing the sugar content requires very little product reformulation, but in others there are significant changes that have to be made in order to produce a palatable product, or one with an equivalent shelf life - sugar can play an important role in preservation (in preventing or delaying microbial growth).

Soft drinks and squashes have traditionally contained sugar as an added sweetener; although the fruit ingredients also contain sugars, reformulating with no added sugar requires the addition of artificial sweeteners to provide an acceptable taste.

With jams, jellies and marmalades, reducing the added sugar content results in a product with significantly less microbiological stability than normal-sugar versions. Jams have long been treated as products that are stable at ambient temperature, even after opening (i.e. they do not require refrigeration). To maintain this level of stability when the sugar level is significantly reduced would require the addition of preservatives - sorbate, benzoate and the bisulphites are all permitted in the EU. Alternatively, depending on the formulation, the product may be microbiologically stable if kept refrigerated.

In other products, sucrose actually contributes to the textural characteristics and final overall eating quality. In work carried out at CCFRA (Anstis and Cauvain, 1998), the use of alternatives to sucrose in cake formulations was investigated. Many of these were chosen either because they were sweeter than sucrose (and so could deliver the same sweetness in a lower volume), or because they had an inherently reduced calorific value (i.e. they were potentially 'less fattening'). Amongst the sucrose alternatives investigated were fructose, glycerol, various sugar alcohols, Litesse (a low-calorie polydextrose [polyglucose] bulking agent), and isomalt (a low-calorie disaccharide). All alternatives resulted in cakes with different characteristics than the standard sucrose cake. Among the characteristics that changed were cake crumb colour, cake volume, moisture content and loss, crumb cohesiveness, hardness and 'recovery' (springiness), and shelf-life (due to more rapid staling or the growth of mould). This is just one example of the complexity of food chemistry and the importance of chemical interactions in producing a nutritious and desirable product.

Starch is broken down in the body by various amylase enzymes. The end result is glucose, which is absorbed through the gut wall into the bloodstream. The glucose is then used as an energy source as described above.

Another form in which we acquire a significant amount of carbohydrate is sucrose (table sugar), which is a disaccharide made up of one glucose molecule and one fructose molecule. Although the metabolism of sucrose will result in roughly the same end products as the metabolism of starch, the latter is generally the preferred form of carbohydrate intake. There are many aspects to this - but there are two main points. Firstly, starch, being a large molecule, is broken down over a relatively long period of time. This generally provides a slow release of sugar for energy generation, rather than a rapid and large burst of energy that would result from eating a similar quantity of sugar. (There are many factors which affect the rate of sugar absorption into the bloodstream and the resultant change in blood glucose levels. The overall characteristic of the food in this respect is termed its glycaemic index). Secondly, starch-containing foods are also usually associated with other nutrients, such as proteins, vitamins, and minerals. Thus, starch intake is associated with a better balanced diet, and a higher volume of food, whereas sugar is usually ingested in pure form or added as an extra - this may result in excess energy-providing material being taken in, which may eventually result in excessive weight gain. Of course, carbohydrate intake in the form of simple sugars can occur without added sucrose. As stated above, fruits sometimes contain high levels of sugars (both sucrose and various monosaccharides) and often no starch, but they are not often eaten in the volumes that would provide all of our carbohydrate needs.

The principal way in which babies take in carbohydrate is as lactose ('milk sugar'), which is a disaccharide consisting of one molecule of glucose and one molecule of galactose. In breast milk, this comes as part of a complete meal, containing all of the fats, proteins, minerals and vitamins required for early life development. We have now come to use cow's milk and other animal milks in various forms (cheese, yoghurt etc). However, for the majority of the world's population, by adulthood, the enzyme responsible for lactose metabolism to glucose and galactose (lactase) declines to about 5-10% of its level at birth. Only about 30% retain enough enzyme to utilise lactose efficiently throughout adult life (predominantly caucasians and northern Europeans). The significance of this to the food industry is discussed later in this book (see Section 4.2).

Other carbohydrates, such as cellulose, we are unable to digest. These will be discussed in the chapter on dietary fibre.

In the body, we can store a limited amount of carbohydrate in the liver and in the muscles - this is stored in the form of another polysaccharide - glycogen. Like starch, this is made up of chains of glucose molecules, which are highly branched. This permits rapid breakdown by enzymes, as enzymes 'nibble away' at the end of the chains and branching provides more 'ends'. The body needs to be able to control the level of glucose in the blood. In simple terms, the hormones insulin and glucagon act in opposition, stimulating either blood glucose conversion to glycogen or glycogen breakdown to glucose. The actual chemical conversion is carried out by enzymes, but the hormones stimulate the production of the relevant enzymes, thus switching the system on and off. If this system malfunctions (the inability to produce functional insulin is the most common reason), diabetes results. People with diabetes need to closely control the amount and rate of carbohydrate intake - especially sugar intake. This, and its significance for food manufacture, is discussed further in a later chapter (see Section 4.4).

The amount of glycogen we can store is limited - excess sugars are channelled into other metabolic pathways, which result in fat synthesis. Other animals have similar metabolic systems, and meat products are generally thus a poor source of carbohydrate. This has great significance in animal husbandry: if the diet of the animal is not carefully controlled, the animal can store any excess as fat, which can make it less desirable as a food source.

2.2 Lipids

Fats or lipids are the second important source of energy in the diet. The term lipid includes both fats and oils - the former being solid and the latter liquid at room temperature. On a weight-for-weight basis they yield over twice as much energy as carbohydrates. There are many different types of lipid, and they are utilised in many different ways by the body – contributing to cell, tissue and organ structures, as well as being a major energy provider. Although they have a poor reputation in terms of the diet, they come in many different forms and they are an essential part of the diet.

Their diversity of roles in the body is their main dietary difference from carbohydrates, which are primarily used as an energy source. In certain cases, the body cannot make individual lipids itself and is totally reliant on a dietary source.

2.2.1 Triglycerides and fatty acids

The bulk of the lipids that we consume and the major energy-providing forms are the triglycerides. These consist of two components: glycerol joined (esterified) to three fatty acids - long-chain molecules with acid ('carboxyl') end-groups. These two components can become separated in food (hydrolysed). If this happens in the food it can result in the formation of off-flavours (hydrolytic rancidity). This hydrolysis itself is of no dietary significance, as it is the first thing that the body does with triglycerides.

The different chain lengths (i.e. number of carbon atoms), and the number and position of double bonds along the chain mean that there are around 15-20 fatty acids regularly occurring in foods.

The fatty acid portion is the major provider of energy, each carbon and hydrogen atom in the chain being combined with oxygen (obtained from the air) to yield carbon dioxide, water and energy. Excess fat is stored as such: although carbohydrates can be converted to fat, there is no return route. However, the role of stored fat cannot be over-emphasised. It is the main form in which we store energy, and it acts as a reserve in times of food shortage or high activity. As all animals store food as fat, animal products are a major source of dietary fat.

However, plants also yield many valuable lipids, including several types of 'unsaturated' fatty acids (i.e. those with one or more double bonds in the carbon chain). Some of these can only be made by plants, and yet are essential for adequate body metabolism. Therefore we need to consume them. Two fatty acids that the body cannot make are linoleic and alpha-linolenic acid: these are the 'parent' acids of the omega-6 and omega-3 fatty acid group respectively. The terms omega-3 and omega-6 derive from one of the nomenclatures used to number the carbon atoms in the fatty acid chain, the omega carbon being the one furthest from the acid end of

Box 5 - Diagrams of triglyceride and fatty acid structures

The carbon atom has a valency of 4, which means that it has 4 'arms' with which to bind to other atoms. In a triglyceride in which all of the available 'arms' of the carbon atoms in the fatty acid portion are linked either to neighbouring carbon atoms or to hydrogen atoms (i.e. a saturated fatty acid), the following structure is obtained, where 'x', 'y' and 'z' may indicate the same or different numbers of carbon atoms (usually 14, 16, 18 or 20):

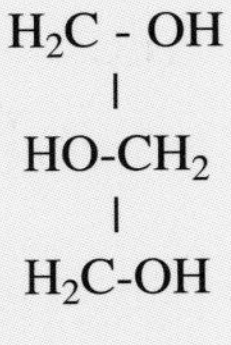

H_2C - OH
|
HO-CH_2
|
H_2C-OH

Glycerol

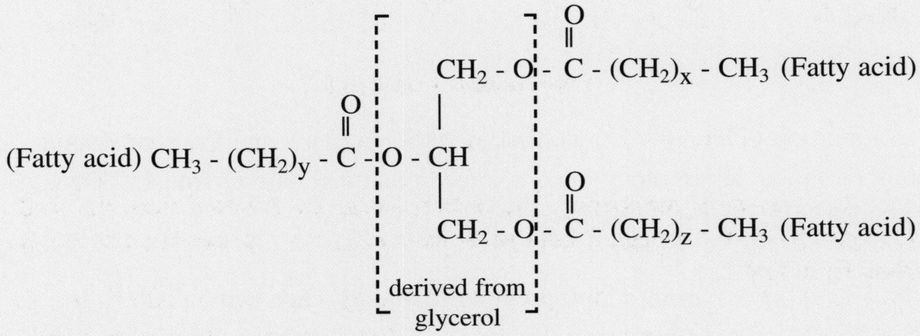

$$
\begin{array}{c}
 O \\
 \| \\
CH_2 - O - C - (CH_2)_x - CH_3 \text{ (Fatty acid)}
\end{array}
$$

$$
\text{(Fatty acid) } CH_3 - (CH_2)_y - \overset{O}{\overset{\|}{C}} - O - CH
$$

$$
\begin{array}{c}
 O \\
 \| \\
CH_2 - O - C - (CH_2)_z - CH_3 \text{ (Fatty acid)}
\end{array}
$$

derived from glycerol

Fatty acids themselves are usually depicted as angled chains, with each angle depicting a carbon atom (and its associated hydrogen atoms) - palmitic acid is shown. In esterification to form a triglyceride, the OH's from each of the fatty acids and the H's from glycerol OH's are lost as 3 water molecules.

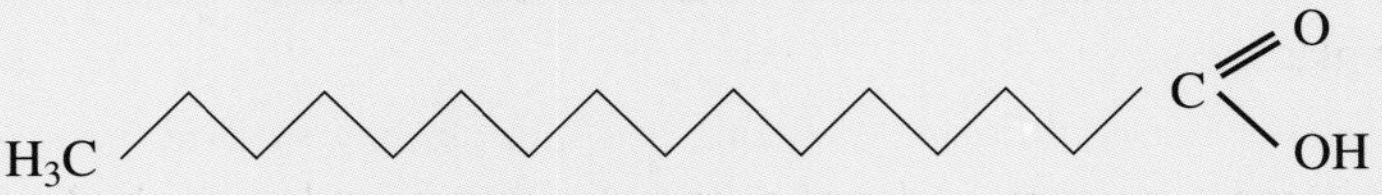

$$
H_3C \diagup\!\!\diagdown\!\!\diagup\!\!\diagdown\!\!\diagup\!\!\diagdown\!\!\diagup\!\!\diagdown\!\!\diagup\!\!\diagdown\!\!\diagup \quad C \overset{O}{\underset{OH}{<}}
$$

Carbon has the ability to make double bonds with itself, with two hydrogen atoms being excluded. Fatty acid chains with one double bond are termed monounsaturated, while those with two or more double bonds are polyunsaturated.

Box 6 - Diagrams of linoleic and linolenic acids

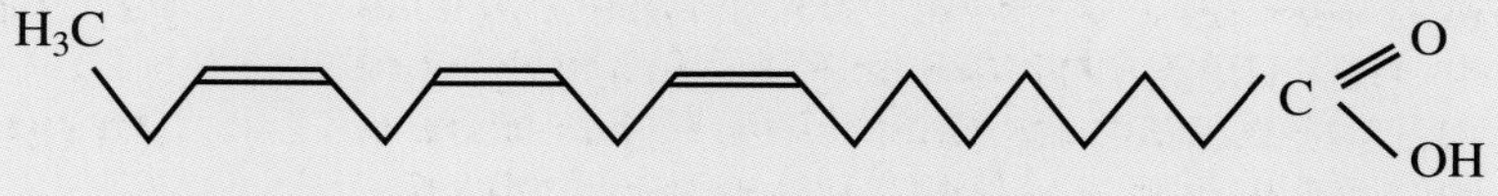

linolenic acid - omega-3

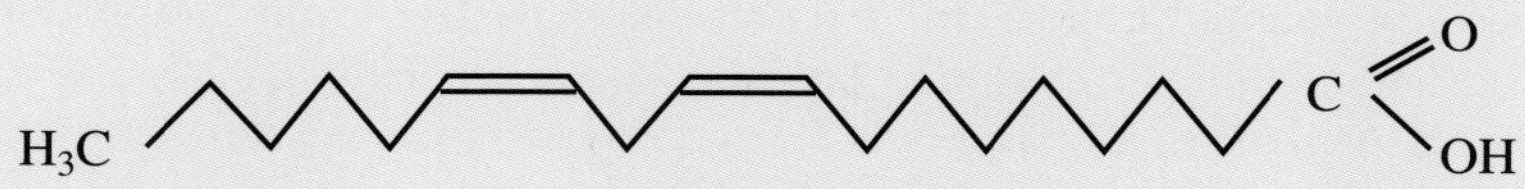

linoleic acid - omega-6

Under this system of numbering, the carbon atoms are counted from the methyl (H₃C) group - so the omega-3 fatty acids have their first double bond at the third carbon from that end.

the molecule. In simple terms, our bodies can only dehydrogenate fatty acids (i.e. form double-bond structures) at certain parts of the chain. In order to get fatty acids with double bonds further down the chain, we need to eat them with these bonds already in place.

As well as being a major provider of energy, fat stores insulate the body and act as shock absorbers - cushioning the vital organs. Fat also helps in the metabolism of proteins and carbohydrates.

Box 7 - Food processing - fatty acid hydrogenation

The variety of fatty acids and triglycerides occurring in foods provides an excellent opportunity to the food industry in food formulation. The triglycerides with shorter-chain length, unsaturated fatty acids exist as light oils, whereas those with longer-chain saturated fatty acids are solid fats and lards. By mixing different ingredients, a range of products can be derived with very specific textural and functional properties. Chemical hydrogenation (breaking the double bond and adding hydrogen atoms) can also be employed to produce saturated fatty acids from unsaturated starting materials. This is useful, for example, if trying to produce a vegetarian product that requires saturated fats in its formulation, as these are more common in animal products. However, hydrogenation is rarely taken to the extreme of all fatty acids being converted from unsaturated to totally saturated. Some mono-unsaturated fatty acids will remain, and some of these will be in the 'trans' form, rather than the 'cis' form, i.e. the hydrogen atoms will be on the opposite side of the double bond, rather than the same side, as shown in the diagram. Trans fatty acids have been linked with heart disease in recent years, and although the exact relationship is not clear-cut, the food industry has responded to consumer concerns by trying to find alternative ways of achieving the correct blend of fats in products to achieve the desired result. It should be pointed out, however, that trans fatty acids do occur naturally in some foods, notably milk and butter.

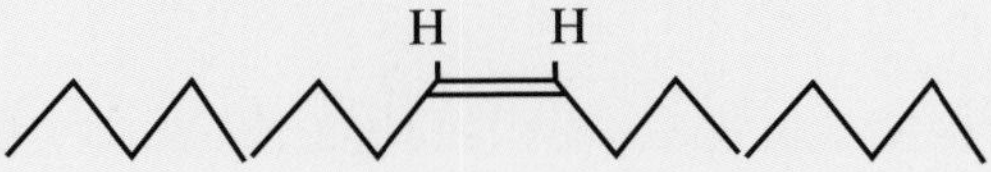

cis

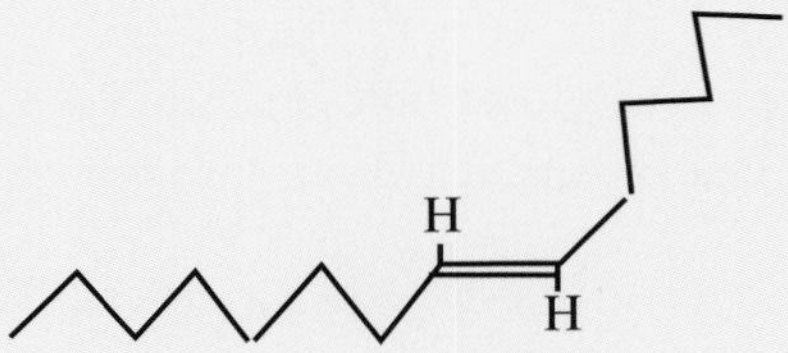

trans

Box 8 - Modifying fat intake

The need to change our dietary habits with regard to fat intake has probably been the most important dietary recommendation from government and health advisors over the last 20 years. However, as will be appreciated, the types of fats in the diet and their various roles in the maintenance of health means that a straightforward reduction in fat intake is not the only step that needs to be taken. There has been much written about the benefit of monounsaturated and polyunsaturated fats in comparison with (mainly animal) saturated fats, the benefit of trying to increase blood high-density lipoproteins at the expense of low-density lipoproteins, and the need to reduce cholesterol levels in the body. However, other valuable nutrients, including vitamins, are naturally associated with fats in primary products, and thus it could be argued that it is in formulated foods that fat intake should be modified.

In response to consumer demand for reduced fat intake, and for a change from animal fats to less saturated plant fats, the food industry has taken many steps to reformulate products and to offer alternatives. Probably one of the first of these, over thirty years ago, was the major expansion in the UK of vegetable-based cooking/frying oils as an alternative to lard. This has continued, and today there is a wide array of oils and blends available.

The dairy industry has been very successful in the development of low-fat products. However, reformulation of other product types has proved more difficult. One of the major users of fats in product formulations is the bakery industry. Reducing fat content is not straightforward, as it plays a major part in mouthfeel, texture and flavour characteristics of the product. In many cases, such as pastry, it is fundamental to the formation of the product, and the quality of the product is often correlated with its fat content.

Box 9 - Fat substitutes and mimetics

The ways in which fat levels can be reduced will vary from product to product. In some areas it has so far proved impossible to produce products that are acceptable from a quality point of view and commercially viable. However, there are many products in which fat levels have been reduced by the use of substitutes or mimetics (i.e. components which mimic the behaviour of fats), such as low-fat spreads.

Traditional ways of reducing fat levels in formulations have included using emulsifiers, substitution by air or even water, using reduced-fat ingredients (e.g. skimmed milk in place of full-fat milk or cream), and baking products instead of frying them. However, most of these operations have limited applications, and much work has gone into developing fat substitutes and mimetics.

Several types of sucrose polyesters have been developed. These consist of a central sucrose molecule esterified to 6-8 fatty acid units. By modifying the fatty acid chain length and degree of saturation, they can potentially be tailor-made to simulate fat in baked and fried goods. They are generally stable at cooking, frying and baking temperatures, which makes them technically useful for bakery products. As the 'outside' of the molecule is fat-based, they function as fats in products, but their large molecular size means that can not be digested and absorbed into the body (they are too big for the digestive enzymes to act upon) and so contribute no calories to the diet. Examples of these include Olestra (from Procter & Gamble) and Salatrim (developed by Nabisco). Unfortunately, from a technology point of view, these types of products have not gained widespread regulatory approval, because of lingering health concerns. Sorbestrin, from Cultor Food Science, is based on the polyol sorbitol, again esterified to several fatty acids.

Emulsifiers can be used in relatively small amounts to replace fats, or to allow less fat in a product to perform the same technological function. Work at the Flour Milling and Baking Research Association (Cauvain *et al*, 1988) showed that glycerol monostearate could allow the fat level in a high-ratio cake formulation to be reduced from 18 to 9% of the cake weight without any change in eating quality being detected. However, these emulsifiers are themselves fats and so care has to be taken in making 'reduced-fat' claims.

There are many fat mimetics available on the marketplace, including over 40 based on starch. These form a gel which holds water and mimics the textural characteristics of fat. They are used to build solids and viscosity, bind and control water, and contribute to a smooth mouthfeel in fat-replacing systems. They have to be able to withstand food processing operations and still provide the required texture in the final product. Native starches from corn or wheat can be used as fat replacers in biscuits and cakes, whereas pre-gelatinised, modified, high-amylose starch is suitable in baked goods, icings and fillings. As they are used to achieve functional and sensory properties, they perform best in higher moisture foods such as cakes, but less well in biscuits and crackers.

Other fat mimetics that have been developed and utilised include maltodextrins, protein-based products and fibre/cellulose-rich ingredients. The latter bind water within their structure, helping to give the impression of fat. Oatrim is one example; it is made by partial hydrolysis of an oat or cornflour bran fraction and can be added as a dry powder or a hydrated gel. It gives a buttery mouthfeel to the final product. Protein-based mimetics include Simplesse (from NutraSweet Kelco), a microparticulated protein. These are formed by hydrating egg white and/or milk proteins and subjecting the mixture to a heating and blending process. Further processing blends and shears the gel to form microscopic, coagulated, deformable particles that mimic the mouthfeel and texture of fat. Although many protein-based fat mimetics are not heat-stable enough to withstand frying, they are suitable for dairy products, salad dressings and products that may be cooked.

There is no single fat replacer that fits all applications. Fat is a complex, highly functional ingredient, and can only be replaced by a combination of other ingredients (Catterall, 2001).

2.2.2 Phospholipids

Phospholipids, although only making up a relatively small portion of dietary fats, have an important role to play in the body. They are structurally quite similar to the triglycerides, but have one fatty acid chain replaced by a phosphate group linked to a nitrogen-containing molecule, such as choline or ethanolamine. The phospholipids containing choline are known as phosphatidylcholines or lecithins.

The chemical structure of phospholipids means that they can dissolve in both water and fat. The body uses this property by incorporating them into cell membrane structures, to facilitate the transfer of fat-soluble chemicals (e.g. some vitamins and hormones) in and out of the cell. In processed foods, phospholipids are often incorporated for their emulsifying properties. Among the richest sources of phospholipids are eggs, liver, soya beans, wheat germ and peanuts (Whitney *et al*, 1998). However, the liver is able to synthesize all the phospholipids the body needs, provided that it has enough 'starting materials'.

Box 10 - Reduced fat bakery products

Recent work at CCFRA (summarised in McEwan and Sharp, 1999) has demonstrated the problems associated with trying to reformulate bakery products with a reduced fat content. Although there are a number of replacers and alternative systems available, for many products it has proved both technically difficult and expensive to produce acceptable products. A survey of manufacturers indicated that there were insufficient ingredients available to produce the new products and that eating quality and shelf-life of what was currently being produced was poorer than for 'standard-fat' recipe products. A lack of knowledge of the technical functionality of alternative ingredients was a major problem. Whilst small reductions in fat content were feasible in some cases, the need for a 25% reduction before a 'reduced fat' claim could be made was often too much, and thus one economic justification was lost. Alternative ingredients were generally more expensive and manufacturers found that not enough customers were willing to pay the extra to make new product development worthwhile.

A survey of consumers suggested that they perceived bakery items as treats and indulgences, often for special occasions, and that small reductions in fat content were not seen as any great benefit. There was also the perception, often from past experience, that the lower-fat items were of poorer eating quality, and that they were more expensive. One solution suggested for this was for completely new lower-fat products to be developed that formed a category of their own, rather than being newer versions of existing products.

2.2.3 Sterols

Both animal and plant foods contain sterols, but cholesterol is found almost exclusively in animal foods. Cholesterol is required for a variety of roles in the body. Its rigid and highly hydrophobic (water-repelling) properties are important for its role in cell membranes and nerve tissues. Many hormones are steroidal in nature, and cholesterol is also the precursor for the bile acids, which act as specific detergents in many metabolic processes (Sadler et al, 1999). However, cholesterol is not required in the diet, as sufficient quantities can be synthesized from other nutrients in the diet. The liver synthesises about 800-1500mg cholesterol per day, which is generally significantly more than is contributed by the diet (Whitney et al., 1998). The level of cholesterol in the body is also subject to close metabolic control, irrespective of the amount ingested.

The dietary significance of phytosterols (plant sterols) has not been well characterised, although one role that is known about is the conversion of ergosterol (which is widely distributed in plant foods) to vitamin D by the action of ultra-violet light. There are several different phytosterols and they may play different roles in the diet. They are believed to interfere with the absorption of dietary cholesterol, and also to inhibit cell reproduction in the gastrointestinal tract, helping to prevent colon cancer.

2.3 Proteins

Proteins can be used as an energy source for the body, if required. However, their primary function is in the formation of body structures, such as muscle. Proteins are made up of amino acids, of which there are around 20 different types in all. The body requires these in various quantities and cannot easily store them. Those surplus to immediate needs are excreted, and a shortage of any one particular amino acid will mean that building work on a particular protein is halted. Therefore it is important to eat a mixture of proteins that will provide enough of each of the 20 individual amino acids.

Amino acids are all based on the common structure shown in Box 11, but they vary considerably in their chemical properties because of the differences in the R groups.

The body can make many of the amino acids, but nine of them (histidine, isoleucine, leucine, lysine, methionine, phenylalanine, threonine, tryptophan and valine) either cannot be made at all or not in sufficient quantities for the body's needs. These are termed 'essential' amino acids, and must be provided from the diet. Others may also be required from the diet in certain circumstances, e.g. tyrosine, which is normally made from phenylalanine (Whitney et al, 1998).

In general, individual animal products have a better balance of amino acids for our needs than plant products. This is not surprising, bearing in mind the use that the animal would have made of the proteins and individual amino acids would be similar to our needs. However, we do not usually eat single products, and an adequate balance of amino acids can also be obtained by eating a mixture of plant

Box 11 - Generalized structure of an amino acid and protein chain

$$H$$
$$|$$
$$R\text{-}C\text{-}NH_2$$
$$|$$
$$COOH$$

where R is one of about 20 different chemical groupings

These individual amino acids are linked to each other through the amino (NH_2) and acid groups (COOH), thus:

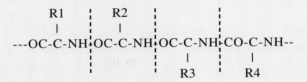

where R1, R2, R3 and R4 may be the same or different chemical groupings

products in a vegetarian or vegan diet. Some plant foods contain relatively low amounts of certain amino acids, and this must be compensated for by eating other products which provide enough of the amino acid in question, as the body requires adequate amounts of all the amino acids to be available at the same time. As stated above, some amino acids can be synthesised in the liver, if the total protein and energy intake is sufficient, but adequate quantities of the essential amino acids have to be obtained from the diet.

Muscle foods such as meat and fish are typical high-protein animal products. Raw chicken meat, for example, contains approximately 20% protein and 75% water. Among the plant sources that are high in protein are beans. Soya beans (14% protein) are particularly widely used (see Box 12).

Box 12 - Soya protein and products

The high level of protein in soya, and its almost perfect balance of amino acids, has made it the most utilised vegetable protein source in formulated foods. Of about 14,000 new food products identified in 1999 and 2000, soya products are listed on the label of over 2200. Ironically, the initial growth of the soybean industry in the West was primarily influenced by its oil production potential rather than its protein content. The latter was not properly recognised until the 1940s. Initial oil extraction methods involved crushing and hot pressing, which resulted in a denatured protein residue of little commercial use to the food industry. The advent of solvent extraction of oil in the 1930s meant that a functional protein fraction by-product was available in large quantities, and commercial production of edible soy protein began in the late 1950s. Earliest food uses were as full-fat flours and grits as substitutes for significant portions of wheat flour in bakery products. Nowadays soy protein products include defatted soy flakes, soy meal, and texturised protein products. The latter can be produced from soy flour and concentrates by thermoplastic extrusion to impart a meat-like texture to the products. This is achieved by mixing with water and additives to form a dough and extruding under high temperature and pressure to obtain the fibrous texture. A spinning process can also be used in which a soy isolate is solubilised in alkali and forced through a spinnerette into an acid bath to coagulate the proteins. The fibres formed can be combined into bundles.

In 1970, a meat-like soy protein product was marketed by Archer Daniels Midland under the trade name TVP (textured vegetable protein). The use of extrusion technology led to continuous slabs of texturised proteins suitable for cutting into various shapes and sizes, resulting in ham, beef and poultry-like products. The proteins are produced from a defatted soy flour, and so are themselves low in fats. However, final products often have significant amounts of fat added back to give succulence and other eating quality attributes to the product.

One of the main reasons that soybeans are so widely used as a protein source is the high 'quality' of the protein, in that its essential amino acid composition is very close to usual human requirements. It is slightly deficient in the two sulphur-containing amino acids, cysteine and methionine, but is rich in lysine. In contrast, cereal proteins have limited lysine levels, but are rich in cysteine and methionine. Thus, the two complement each other nutritionally very well (Liu, 1997).

Box 13 - The development of Quorn mycoprotein

Quorn is the registered trademark of a mycoprotein-based family of products. In the 1960s, Ranks Hovis McDougall (RHM) put a great deal of effort into identifying a 'new' food source to try and help address a potential world food shortage that was envisaged by many at the time. A fungus, *Fusarium graminearum*, was identified near wheat fields in Marlow that showed promise as a significant source of protein (mycoprotein = fungal protein). Work concentrated on determining whether it would be safe for human consumption and, in a joint venture with ICI, whether it could be produced via fermentation technology in large enough quantities and cheaply enough to be commercially viable (Wilson, 2001).

In 1984, MAFF granted approval for its use as a human food (Sadler, 1988) and the first Quorn product - Savoury Pie - was launched through J. Sainsbury in the following year. In the late 1980s, a joint venture between RHM and ICI was created (Marlow Foods) to commercially develop the new food. Products were introduced into Belgium in 1991 and the Netherlands and Germany in 1992 (Sharp, 1994), but it was not until 1993 that products became available nationally in the UK and across Western Europe. Approval of the mycoprotein in the US and the subsequent launch of products occurred at the end of 2001.

The acceptance and development of Quorn from a consumer's point of view (in addition to the technical and commercial feasibility issues mentioned above) has depended on two main characteristics: its good nutritional profile, and the ability to produce foods with a good flavour and texture profile from it. Quorn is relatively high in protein (12.2%) and fibre (5%), and low in total and saturated fat (2.9 and 0.6% respectively). The protein has a high biological value - i.e. it has close to the ideal combination of amino acid levels for our dietary needs (similar to the milk protein, casein), the limiting amino acids being methionine and cysteine. Fibre levels are higher than in most fresh vegetables, with chitin and beta-glucan cell wall material predominating. Quorn contains most of the B vitamins, with the exception of B12, which is absent. It is particularly high in biotin, which is incorporated into the fermentation broth to ensure optimal growth.

In the formulation of products, the Quorn is mixed with egg albumen and flavours and processed into the desired texture (different processes will result in different types of texture to suit the end product). As Quorn has little flavour itself, other desired flavours can be added, again to suit the end product.

The number of Quorn products available has increased dramatically in the past 7 years. Worldwide turnover in 1993 was less than $3 million; in 2001, it was estimated that it would exceed $150 million.

As well as dietary significance, proteins play a significant role in the formulation of some foods. One of the most notable examples is in the production of bread doughs. The gluten proteins from wheat are a key component in the development of dough structure. The breeding of wheat varieties with suitable levels of gluten, and the blending of flours to get just the right balance for a specific application, has been a major part of the bakery industry for hundreds of years. Formulation of high quality products without wheat gluten is very difficult. As will be discussed in a later chapter, this is highly significant to those people suffering from coeliac disease (see Section 4.1), which is an intolerance to wheat gluten and similar proteins.

2.4 Other energy-yielding components

In addition to proteins, fats and carbohydrates, there are three other classes of compounds routinely found or used in foods that yield energy in the diet: polyols, organic acids, and ethanol. Although they do not usually contribute significant amounts in most foods, they are required by law to be included in the calculations of the total energy content of a food.

2.4.1 Polyols

Polyol is a term used to describe a chemical with alcohol groups on several carbon atoms. The most widespread in nature is glycerol, which is usually esterified to fatty acids in triglycerides and phospholipids. Unless present in a food as 'free' glycerol (i.e. not in combination with fatty acids), this would be covered in declarations of the fat content of the food (see above). Most of the other polyols occurring in foods are sugar alcohols: sorbitol, mannitol, xylitol, maltitol and lactitol, derived from the sugars glucose, mannose, xylose, maltose and lactose respectively. In addition, isomalt is a mixture of glucose-sorbitol and glucose-mannitol compounds (Smith, 1991). Sorbitol is found naturally in some fruits, and mannitol and xylitol are also naturally occurring, but most dietary intake of polyols occurs as a result of their use as sweetening agents and bulking agents. Although they are generally slightly less sweet than glucose and sucrose, they have low cariogenicity, their metabolism is insulin-independent and they do not cause a rise in blood glucose levels after consumption. Polyols are used in confectionery products as bulking agents, i.e. at levels above 10% - much higher than those needed to sweeten the product. Intakes of 20-80g/day (depending on the polyol in question) can result in laxative effects, and it is a legislative requirement to indicate this on the packaging of food containing more than 10% added polyol.

2.4.2 Organic acids

Many organic acids occur in small amounts in a wide variety of foods, and several of these are also used as acidulants in food formulations. These include malic, ascorbic, citric and tartaric acids. Malic acid is found in relatively high levels in many fruits such as apples (0.4%), quince (0.9%), and sweet and morello cherries (0.9 and 1.8% respectively), as well as plums, peaches, apricots, raspberries and strawberries. Citric acid is also widely distributed, and particularly high levels occur in passion fruit (3.25%), blackcurrants (2.4%), citrus fruits (1-5%), strawberries (0.75%) and gooseberries (0.75%). Tartaric acid is found in rose hips (1.65%), while quinic acid is a major acid in kiwi fruit (1%). Ascorbic acid (vitamin C) occurs in low levels in many fruits, but blackcurrants and guava are particularly good sources (about 0.2%)

As well as these fruit acids, two other organic acids occur in significant quantities in specific foods: acetic acid in vinegar and similar fermented products, and lactic acid in fermented milk products such as yoghurt and cheese (typically up to 1%) (Scherz and Senser, 1994).

However, the amount of energy these acids contribute to the diet is relatively small, overall.

2.4.3 Ethanol

Ethanol is the end-product of yeast fermentation in alcoholic beverages. Levels in beers and lagers are typically 3-5%, whilst in wines and spirits levels can range from 10-30%. In most cases, intake is limited by its side-effects. However, ethanol is almost as energy rich as fats from a dietary point of view, and its contribution to calorie intake can be significant.

2.5 Fibre

Fibre is not a nutrient in the same way as starch or protein: it does not provide energy and is not involved in energy metabolism at the cellular level. However, it does influence many digestive processes and so is included here to indicate its relevance to human nutrition. The functional fibre in our diet is derived from the macromolecular chemicals in plants (including algae and fungi such as the Quorn mycoprotein fungus) that contribute to its structure. The main characteristic of fibre from a dietary point of view is that it cannot be digested by human digestive enzymes, although some can be broken down by gut bacteria. Fibre plays a major role in for the maintenance of correct function in the gut. There is also evidence that certain types of fibre can affect the way that fat is transported in the bloodstream.

Most dietary fibre groups are polysaccharides of one type or another. Celluloses are the main constituent of plant cell walls, and therefore occur in all plant-containing foods. Hemicelluloses are a major constituent of cereal fibres; they are a very diverse group of chemicals, some of which are soluble. Pectins are widespread in vegetables and fruit, especially citrus fruits and apples. As well as being consumed as the normal constituents of these foods, they are widely used as ingredients by food manufacturers as thickeners and gelling agents. Gums are another group of dietary fibre chemicals that are widely used in the food industry as additives, having stabilizing and related properties. Examples are guar gum, carrageenan, gum arabic (acacia), karaya gum, agar, xanthan gum, tragacanth and locust bean gum. Lignin is a non-polysaccharide fibre and is the characteristic constituent of woody tissue. Because of its texture it is not usually a major edible part of foodstuffs, although it is a significant component of dried fruits

Because fibres are such a disparate group of chemicals they have many different functions in the body, and in some cases one type will have a directly opposing effect to another type (such as in delaying or accelerating food transit through the intestine). Most delay glucose absorption, and the soluble fibres (found in citrus fruits and apples, oats, barley and legumes) are believed to lower blood cholesterol.

The food manufacturing industry has taken advantage of some of these beneficial effects by the creation of several new product lines. Among these have been oat-based cereal bars, and bakery products fortified with psyllium, a herb-derived soluble fibre.

Box 14 - Measuring dietary fibre

Because dietary fibre comprises such a wide range of chemical macromolecules, different analytical methods will give different values for the amount actually present in food. Having different values does not mean that one is more correct than another, it merely reflects what the method defines as fibre, and what it is actually capable of measuring. Analysis of complex macromolecules is not straightforward and relies on various assumptions and approximations being made. In the case of dietary fibre, this is further complicated by the many different types of molecules that are being analysed.

With increasing interest being focused on the importance of dietary fibre, and the number of products being formulated and marketed with increased dietary fibre content, it is important that the figures given in nutrition declarations are directly comparable with each other, and that claims that a product is high in fibre, for example, are measured against a set standard. In McCance and Widdowson's Composition of Foods (at least until the 5th edition - Holland *et al,* 1991), the main UK publication on food nutritional composition, results from two methods (Southgate and Englyst) are quoted for dietary fibre. These sometimes give significantly different answers, with the Southgate method giving generally higher figures. The Englyst method has been the standard method used in the UK for many years, and fibre labelling and advertising claims have had to be based on these figures (see Food Labelling section). Recently, the UK decided to standardise on the Prosky method - also known as the AOAC (Association of Official Analytical Chemists) method, which is the standard method in the USA and in much of mainland Europe.

The Englyst method predominantly measures non-starch polysaccharides (inulin, cellulose, hemicellulose, pectins etc), whilst the AOAC method additionally measures some lignin and some resistant starch. The latter is starch that escapes digestion in the small intestine, either because it is physically inaccessible, or is present in resistant granules, or has retrograded. (This is a structural change, particularly in the amylose component, that occurs when the starch recrystallises after cooking and resultant solubilisation.) It is retrograded starch that the AOAC method measures.

On average, the AOAC method gives fibre values for foods about 1.4 times higher than the Englyst method, but there is much variation from one food to another, depending on the type of fibre it contains. There is currently much debate as to whether the criteria for making a fibre claim (i.e. the measured amount of fibre in a food) should be altered to compensate for this change. However, it seems likely that the current guidelines will remain largely unaltered.

2.6 Minerals

The body needs a myriad of elemental minerals for correct function. Among these are potassium, sodium, calcium, zinc, phosphorus, magnesium, iron, manganese, selenium, and iodine. Zinc, for example, has structural, regulatory and catalytic roles in a number of enzymes, and also has a structural role in non-enzymic proteins. A deficiency of zinc can result in growth retardation and defects in the skin, intestinal mucosa and immune system (DoH, 1991b). Iron is a significant component of haemoglobin, the red pigment in blood cells responsible for transporting oxygen to all cells in the body, and a deficiency will result in anaemia, with tiredness and general lethargy being two of the symptoms. Calcium and phosphorus are major components in bones, and both sodium and potassium are required for maintaining ionic balance in cells, and particularly the correct functioning of nerve cells.

Some minerals are required in greater quantities than others and this requirement can vary considerably from one part of the population to another. For example, adolescents and breast-feeding women require more dietary calcium than other adults. The levels of minerals required per day are, generally, in the milligram range.

Unlike some vitamins, minerals are not destroyed by processing, and in some cases their bioavailability (i.e. the ability of the digestive system to extract and absorb them from the food) may be improved. However, they are prone to leaching out of food into cooking water on processing. Thus, avoiding cooking with excess water during commercial and domestic processing can significantly enhance the retention of minerals in processed food. In some cases, it may be that the raw food itself is deficient in a mineral. This may be the case with selenium (see box). In others, there may be other chemicals in the food that prevent it from being available.

Brief details are given in Table 3 of the best food sources of some of the major dietary minerals (see Whitney *et al*, 1998 for further details). For other minerals, although there may be a requirement for them, availability and general intake is such that their dietary source is not a major concern.

Box 15 - Selenium content of foods

Selenium is an element required in small amounts in the diet. It is present in most foods, with Brazil nuts and fish being particularly good sources. The selenium content of plant foods depends to a great extent on how much is present in the soil in which they are grown, and the form in which it is found. Similarly, the selenium content of animal food will depend on levels in feedstuffs. In the UK, concerns about selenium deficiency diseases in farm animals led to animal feeds being supplemented at nutritional levels in 1978. This has virtually eliminated the problem of selenium deficiency and the resulting financial losses due to animals becoming unfit for the human food chain. Processing of food probably has very little effect on the total amount of selenium ingested (BNF, 2001).

Despite the fortification of animal feeds, the estimated intake of selenium in the UK suggests that there may have been a significant decline in the 1990s, from 60µg/day in 1991 to 39µg/day in 1997 (MAFF, 1999), although there is a degree of uncertainty in these calculations. However, these estimated figures compare with the Reference Nutrient Intakes of 60 and 75µg/day for women and men respectively. The Committee on Medical Aspects of Food, however, states that, in relation to this: 'there is, at present, no evidence of adverse health consequences from current intakes...and further research should be encouraged to investigate whether the current levels of intake are adequate, and whether the body adapts to changing intakes'.

Further reading:

British Nutrition Foundation (2001) Selenium and Health. BNF Briefing Paper.

Ministry of Agriculture, Fisheries and Food (1999). 1997 Total diet study: aluminium, arsenic, cadmium, chromium, copper, lead, mercury, nickel, selenium, tin and zinc. Food Surveillance Information Sheet No. 191

Table 3 - Typical dietary sources of minerals

Potassium: fruits (e.g. bananas); vegetables (e.g. potatoes), and to a lesser extent legumes and meat

Calcium: milk and dairy products; tofu; sardines

Sodium: added salt in products

Chloride: added salt in products

Phosphorus: animal protein foods; added phosphates in food

Magnesium: nuts; legumes; whole grains; dark green vegetables; seafood; cocoa and chocolate

Iron: red meat, poultry, fish and shellfish; eggs; legumes; dried fruits; tofu

Zinc: red meat, fish, shellfish (especially oysters) and poultry; whole grains; vegetables

Iodine: table salt (with added iodine); seafood; bread; dairy products. (The sea is the major resource of iodine; plants grown in soil near the sea, and animals fed on these plants are good sources of iodine)

Selenium: seafood; meat; grain

Manganese: widely distributed

Copper: meat; drinking water (depending on the type of pipes used)

Fluoride: drinking water (if fluoridated); tea; seafood

Chromium: liver; yeast; grains; nuts; cheese

Box 16 - Bioavailability of micronutrients

The chemical composition of foods can sometimes give a false picture as to the nutritional value of the food. The efficiency with which any dietary nutrient is used in the body (its bioavailability) is of great significance. For most of the macronutrients, bioavailability is high (90% or more), but for others, particularly vitamins and minerals, it is low and unpredictable. With the micronutrients there are several factors that might make a food that contains a certain vitamin or mineral, not a good source from a nutritional point of view. Alternatively it may be a good source, but only when considering what else is being eaten at the same time.

The bioavailability of a vitamin or mineral can be restricted in three main ways: it may be physically 'locked away' inside the food and not easily released from the tissue of the food; it may be present in a chemical form which cannot be utilised during digestion and cannot be absorbed (or cannot be metabolised after absorption); or it may become bound up during digestion by other components of the diet. Additionally, it may require the presence of other components to facilitate its absorption or availability, and it may not be possible to store large quantities, making rich sources of the chemical no better than moderate sources. Sometimes, these factors work in synergy or opposition to further complicate matters. As an overriding generalisation, absorption efficiency will increase when the nutrient is in short supply in relation to the amount required.

Phytic acid, which occurs in several plant foods, can chelate (entrap) mineral ions such as calcium, iron, copper, zinc and magnesium and prevent their absorption. Oxalic acid is similarly a potential problem with calcium absorption. In foods where levels of these antinutrients are high, an apparantly high level of minerals in the food may be of no dietary consequence, as they will only be poorly absorbed. With calcium in particular, absorption is enhanced by vitamin D, and moderately low levels of calcium in the diet may only be a significant problem if vitamin D is limited.

Vitamin C cannot be stored by the body. Levels in excess of what is required are absorbed but immediately excreted. Excess magnesium is similarly absorbed and excreted, up to 2 g/day. Levels higher than this pass through unabsorbed.

Iron absorption is affected by many factors. It is enhanced by ascorbic, citric, lactic, malic and tartaric acids, fructose, sorbitol, ethanol, and the amino acids cysteine, lysine and histidine, and it is inhibited by tannins and polyphenols (e.g. in tea), phosphates, phytate, bran, lignin and egg albumen, egg yolk and legume proteins, as well as other inorganic elements. Iron is absorbed in the reduced (ferrous) state, rather than the oxidised (ferric) state, which is a major reason why it is enhanced by ascorbic acid (vitamin C - an antioxidant).

2.7 Vitamins

Like minerals, vitamins are required in small amounts. They were originally perceived as minor constituents or activities which could be derived from food and which were essential in small amounts for maintenance of health. There are about 12 major vitamins which the body requires in order to maintain a healthy metabolism. Deficiency in one or more of these vitamins can lead to serious illnesses. Different foods can be good sources of one or more vitamins, but it is important to take into account the fact that some vitamins (especially the B group vitamins and vitamin C) are not completely stable to processing and cooking. Having said this, the widespread occurrence and levels of most vitamins in foods means that deficiency diseases in the Western world are uncommon, despite bioavailability issues and processing and cooking losses.

Vitamin C and the B-group vitamins are water-soluble, whereas vitamins A, D, E and K are fat-soluble.

The concept of vitamins as individual chemicals is actually misleading. The terms were originally developed to describe a biological activity, without knowing what entity was causing it. Thus, while vitamins A and C can be related to retinol and ascorbic acid respectively, vitamin D and E activities are actually the result of more than one chemical. Furthermore, the elucidation of the vitamin B complex of activities resulted in a very mixed situation, with some B vitamins not having a number at all, and some compounds being given different designations as different groups of researchers independently described their effects. Wagner and Folkers (1964) give a short summary of the development of the nomenclature.

2.7.1 Vitamin A

Vitamin A is a fat-soluble vitamin, which is required for growth and for normal development and differentiation of tissues. It is relatively stable and so food processing operations do not present too much of a problem regarding intake. In the retina, the aldehyde form of the vitamin is involved in light-gathering for sight, and one of the earliest signs of vitamin A deficiency is impaired adaptation to low-intensity light (night blindness) (Department of Health, 1991b).

Box 17 - Oxidation and reduction

Oxidation and reduction are two major types of chemical reaction that occur throughout nature, including biological systems. Many of the vitamins that we require are involved in these types of reactions. In simple terms, oxidation is the removal of electrons from a chemical or part of a chemical. In practical terms this often means the addition of oxygen atoms or the removal of hydrogen atoms from the chemical. Reduction is the opposite of this: the removal of oxygen or the addition of hydrogen or electrons. In biological systems these two opposite reactions often occur in tandem and in balance. Many of the B vitamins are involved in the transfer of hydrogen from one chemical to another in energy-generating processes. These are called redox reactions.

The antioxidant vitamins C and E are involved in 'mopping up' oxidative chemicals in the body that might otherwise cause undesirable oxidative reactions. They do this by becoming oxidised themselves, and therefore protecting the vulnerable chemicals. Unless they have a way of being safely reduced back to the active state, levels of the oxidised forms will become depleted, and a regular dietary supply will be required.

Vitamin A can either be acquired in a preformed state (as retinol or retinyl esters), or as carotenoids - almost entirely beta-carotene, which is cleaved to yield two molecules of the vitamin.

Between a quarter and a third of dietary vitamin A is derived from beta-carotene, which is an orange-red pigment found in many vegetables, notably carrots (hence the adage that carrots are good for seeing in the dark). Peppers, brassicas, spinach and peas are also good sources. Among the fruits, oranges, tomatoes, mangoes and melons are high in carotene, and apricots and related fruits are also reasonably good sources.

Pre-formed vitamin A occurs naturally only in animals, the richest sources being liver (the site of vitamin A storage in animals) and dairy products. Although milk itself is not a particularly good source (being only 4% fat), vitamin A is concentrated in fat-based products such as cream and cheeses.

Although vitamin A deficiency is common in some developing countries, it is not generally a problem in the UK, with the Estimated Average Requirement of 400-500 micrograms retinol equivalent/day being easily achieved. Excess vitamin A is stored and degraded, if necessary, by the liver. However, significant excess vitamin A intake, such that the liver can neither store nor degrade it, can result in liver and bone damage, hair loss, double vision and other abnormalities (Department of Health, 1991b). In excess, it is also a teratogen, i.e. it can cause developmental abnormalities in the unborn foetus. Such excess intake is generally the result of vitamin supplement intake, rather than eating too many vitamin A-containing foods.

Box 18 - Genetically modified rice

Rice has been genetically modified to produce high levels of beta-carotene, which could have great significance in combating blindness resulting from lack of vitamin A in Asian countries that rely on rice as a staple food. In a deal widely announced in the media in May 2000, it was reported that farmers in China, India and other rice-dependent Asian countries would be given access to the GM rice, while Astra-Zeneca would commercialise the product in the developed world. Some groups generally opposed to GM developments subsequently gave their support to the Third World aspect of the deal.

2.7.2 Vitamin C

Vitamin C or ascorbic acid has been known for centuries to be required for preventing scurvy and aiding wound healing. That the disease could be prevented by eating citrus fruits was utilised by the navy as long ago as the 1700's, and British sailors were given the nickname 'limeys' by Americans as a result. It has several other roles in the body, including as an antioxidant and in the absorption of iron from the gut. Estimated average requirements are given as about 25 mg/day by the Department of Health (1991b), and the recommended daily allowance (RDA) used for labelling purposes is 60 mg/day. Although citrus fruits are the best-known source of vitamin C, other fruits and many vegetables provide similar or significantly greater levels. Dairy, meat, fish, nut and seed products contain virtually no vitamin C. When calculating the vitamin C 'value' of a fruit or vegetable, the vitamin's stability or lack of stability must be considered. This is a complex issue; not only is a proportion of vitamin C destroyed by heat, it is also susceptible to oxygen (being an antioxidant, one of its main functions is to 'mop up' oxygen or oxidising chemicals in the body). It may also be lost during storage of the raw product. For example, freshly dug potatoes contain around 21mg of Vitamin C per 100g. This falls to 7mg after 9 months' storage (Holland *et al*, 1991). Some typical values for vitamin C contents in raw and cooked fruit and vegetables are given in Table 4.

Although potatoes do not appear to be a rich source of vitamin C, the volume of potatoes that we consume mean that they do make a significant contribution.

Table 4 - Typical vitamin C content of some fruits and vegetables
(Holland *et al*., 1991)

Fruit	Amount (mg/100g)	Vegetable	Amount (mg/100g)
Blackcurrants, raw	200	New potatoes, raw	16
Blackcurrants, stewed	115	New potatoes, boiled	9
Grapefruit, raw	36	Mange-tout peas, raw	54
Guava, raw	230	Mange-tout peas, boiled	28
Guava, canned in syrup	180	Brussels sprouts. Raw	115
Lemons, raw	58	Brussels sprouts, boiled	60
Oranges, raw	54	Red peppers, raw	140
Strawberries, raw	77	Red peppers, boiled	81

2.7.3 Vitamin D

Vitamin D comes in two main forms: ergocalciferol (vitamin D2) and cholecalciferol (vitamin D3). It has a single role in the body - as a precursor for hydroxy- and dihydroxyvitamin D. These are involved in maintaining calcium balance in the body, that is, closely controlling the level of calcium in the blood, by varying the amount that is absorbed from the diet and the amount that is excreted. Calcium's main function in the body is as a major component of teeth and bones.

Box 19 - Vitamin C in orange juice - effects of processing and storage

Although vitamin C is not heat-stable, this is not the only major stability factor of concern to the food industry. Being an anti-oxidant it is more susceptible to the presence of oxygen. Fruit juices are a major source of vitamin C, and much work has been done to maximise the level of the vitamin in the finished product. There are many factors that interact to affect the initial level of vitamin C in the product and its subsequent retention. These include the starting levels in the raw material (which in turn will depend on the variety, growing conditions, age and maturity of the fruit), type and degree of processing, nature of storage container, temperature of storage, and other chemical factors (e.g. preservatives) present in the formulation. The number of factors involved and their interactions makes it difficult to compare different pieces of research that have used different conditions. However, some general trends do emerge.

Heat treatment undoubtedly does destroy some vitamin C activity, as well as affecting other quality factors. Novel processing methods to retain vitamin C have included the use of extremely high pressure (see box in flavours chapter) and pulsed electric fields (PEF) as alternatives to thermal processing.

Ayhan *et al.* (2001) evaluated the effect of different containers on vitamin C stability during storage after PEF. It was found that a processing regime which resulted in an orange juice of good starting quality and extended shelf life had a vitamin C content of just under 60 mg/100 ml. Storage in glass bottles resulted in almost complete retention of the vitamin for 110 days, with polyethylene terephthalate (PET) bottles also giving good retention (approximately 80%). However, high- and low-density polyethylene packaging resulted in around 60 and 80% loss in vitamin over the same storage period. This was ascribed to the greater permeability of these materials to oxygen. Lian *et al* (2000) found that storage in glass bottles in the dark resulted in 62% retention of vitamin C after a year, but that greater loss occurred after storage in the light. Lee and Coates (1999) observed about a 20% decline in vitamin C levels in freshly squeezed unpasteurised orange juice after 2 years' storage at -23°C, whilst Lee *et al.* (1995) showed that elevated temperatures (40-50°C) resulted in very rapid vitamin C loss (87 and 96% in cans and bottles, respectively, after 24 weeks), compared with storage at 30°C (24 and 50% losses, respectively).

continued...

On a practical level, Massaioli and Haddad (1981) found that, in a survey of commercially available orange juices, most remained satisfactory sources of vitamin C through their shelf-life. Figures from the USDA Nutrient Data Laboratory indicate that 30-40 mg/100 ml is a typical value for vitamin C levels in commercial juices when opened. Thus, a 200 ml serving would provide 100% of the RDA. However, it should be noted that vitamin C levels in juices will decline quite quickly after opening the package - decanting into an open contained and leaving unrefrigerated in the light is probably the worst of all options.

Ergocalciferol can be formed from ergosterol (a sterol which is widely distributed in plants) by the action of UV irradiation. Cholecalciferol is similarly formed by the action of the UV in sunlight on 7-dehydrocholesterol in the skin. This is how we obtain the vast majority of our vitamin D. Food provides relatively little, with eggs (5 μg/100g), margarine and spreads (fortified to about 8 μg/100g), and breakfast cereals (typically fortified to 2 μg/100g) being the major sources.

2.7.4 Vitamin E

Like vitamin C, vitamin E's main function is as an antioxidant. It protects lipids and other cellular components from destruction through oxidation. Vitamin E activity actually results from several closely related compounds, the tocopherols and the trienols. However, alpha-tocopherol accounts for about 90% of vitamin E activity in human tissues.

The levels of vitamin E required depend on the level of intake of polyunsaturated fatty acids (PUFA), as these are the main components which the vitamin protects from oxidation - the more PUFA consumed, the greater the requirement for vitamin E. The Department of Health (1991b) reported that daily intakes of 3 and 4 mg of alpha-tocopherol equivalents were adequate for women and men respectively. Whitney et al. (1998) give figures of 8 and 10 mg/day respectively. However, vitamin E is widespread in foods, with cereal and vegetable oils being particularly good sources (although the vitamin is destroyed at frying temperatues), and deficiency is rare.

2.7.5 Vitamin K

The primary role of vitamin K is in helping to maintain the blood-clotting system. Blood-clotting involves a complex chain of reactions, and vitamin K is required for the synthesis of many of the proteins involved in the reaction. As with vitamins D and E, vitamin K is not a single chemical but is derived from phylloquinone, which is found in plants, and menaquinones, which are synthesised by intestinal bacteria. The synthetic menadione is also active. Brassicas and other leafy green vegetables are major sources, as are milk and liver (as vitamin K is stored in the liver). However, about half of our intake derives from that synthesised by intestinal bacteria, and so major impairment of the digestive system is the most likely reason for deficiency, which is very rare.

2.7.6 The B vitamins

The original vitamin B activity was designated as something which combatted beriberi (which literally means 'I can't, I can't' - beri signifying 'weakness'). In fact, this general weakness proved to be a whole range of symptoms, with a variety of causes. The 'factors' that relieved these symptoms were given vitamin designations. Some of these treatments have become accepted as individual vitamins, i.e. related to minor ingredients in food. Others were subsequently found to be due to other factors such as metabolic disorders.

Thiamin (vitamin B1)

Recognition of vitamin B1 activity dates back to 1890, when it was observed that hens fed exclusively on polished rice developed polyneuritis symptoms (disfunctioning of the nervous system) similar to those seen in beriberi patients, who often come from areas where polished rice is a dietary staple. Adding bran to the diet relieved the symptoms (Wagner and Folkers, 1964). This was soon shown to be a nutritional requirement, rather than something needed to combat a toxin in polished rice.

Thiamin requirement is related to the energy requirements of an individual, especially that derived from carbohydrate. This is because one of its main roles is as a co-factor in many of the reactions of energy metabolism.

Box 20 - Food fortification

The level of fortification of foods with vitamins and minerals varies from product to product and from country to country. In the USA, various regulations and guidelines have been put forward based on which vitamins that segments of the population are likely to be potentially deficient in. In the UK, it is a legal requirement to fortify both brown and white flour (but not wholemeal) and margarine, the former with calcium, iron, thiamin and niacin, and the latter with vitamins A and D. There have also been recommendations from various government and expert groups to fortify other products (e.g. to fortify other yellow fats with vitamins A and D to a similar level to that required for margarine) (DoH, 1991a; BNF, 1994), but in general fortification is voluntary.

When fortifying products with nutrients, it is important to ensure that the nutrients survive the processing of the product and storage over its proposed shelf-life, and that they are biologically ' available' (i.e. able to be utilised) upon consumption. It is also important not to adversely affect the flavour, texture and stability of the food itself. The type of claim that the manufacturer may be able to make is also of significance. If the level of a vitamin that can be added is not large enough to allow a claim to be made, then the manufacturer may not deem it worthwhile making the addition, from both a commercial and a nutritional point of view. It is also usually necessary to add more than will be claimed on the label to allow for the losses that will inevitably occur during production and storage. Knowledge of the stability of the nutrients in the products, and how they react with the other ingredients, including each other, is therefore very important.

Some examples of fortified foods available in the UK

Margarine and other Yellow-fat Spreads	Vitamins A and D
Brown and White Flour	Calcium, Iron, Thiamin, Niacin
Breakfast Cereals	Thiamin, Riboflavin, Vitamin B6, Folate, Vitamin D, Iron
Soft Drinks (including Juice Drinks)	Vitamin C
Soya Milk	Vitamin B12
Low-fat Milk Products	Vitamins A and D, Calcium
Vegetable Protein (Meat Analogues)	Thiamin, Riboflavin, Vitamin B12, Iron, Zinc

With the exception of those foods required by legislation to be fortified, non-fortified versions may also be available. There is also a growing market for specific foods to be targeted at sectors of the population with a certain nutritional/health objective in mind - the so-called functional foods or nutraceutical products.

Thiamin can be derived from many foods. Pork and ham are particularly good sources. However, thiamin is destroyed by prolonged cooking and, like other water-soluble vitamins, it can leach into cooking water during boiling or blanching. Therefore, cooking methods that do not use much water (e.g. microwaving) can help to conserve the vitamin. Thiamin is one of the vitamins with which breakfast cereals are fortified. Holland *et al* (1991) give 1mg/100g as a typical figure.

Riboflavin (vitamin B2)

Riboflavin was first recognised as a factor in yeast extract that was required for the maintenance of health, including normal vision and skin health. In fact, the characteristics of this factor proved to be due to a combination of riboflavin and pyridoxol (a member of the vitamin B6 activity complex). Like thiamin, it is involved in energy metabolism, and plays an essential role in all the oxidative processes on which man and other organisms depend. However, its chemical stability is significantly different from that of thiamin. It is heat-stable, but is light-sensitive. Milk and dairy products are good sources, providing about half of the daily intake of the vitamin, with meat and green vegetables providing the rest. Liver and kidney are by far the richest sources of the vitamin per unit weight.

Riboflavin is a vitamin that vegans (who do not eat any animal product) must make a particular effort to consume. In addition to green vegetables, yeast extract products are suitable and excellent sources.

Niacin

This is another vitamin activity that is involved as a coenzyme factor in many reactions in basic cell metabolism. It was originally identified as the factor that prevented the disease pellagra, which is characterised by severe sunburn-like lesions in areas of the body exposed to sunlight, and in areas such as the knees, ankles, wrists and elbows, which are subject to pressure (Department of Health, 1991b). Its activity resulted in the original designation vitamin PP (pellagra prevention).

Niacin activity is in fact due to two distinct compounds: nicotinic acid and nicotinamide. It is unique among the B vitamins in human metabolism in that we can synthesise it from the amino acid tryptophan. Preformed niacin contributes about half of the dietary vitamin requirement, the rest being synthesised by the body from tryptophan. However, most normal diets provide enough tryptophan to meet the need entirely. Diets low in protein can lead to pellagra. This occurred commonly with the widespread use of corn (maize) in the diet, as corn is both low in available niacin and low in protein. In the early 1900's, pellagra in the southern United States resulted in around 10,000 deaths per year.

Niacin is another vitamin that is used to fortify breakfast cereals, to about 18mg/100g. Very high levels of niacin intake (around 3-6g/day) have been shown to cause liver damage.

Vitamin B6

This is another vitamin which is, in fact, an activity provided by a mixture of chemicals: pyridoxal, pyridoxine, pyridoxamine and their phosphates, which are metabolically interconvertible. Pyridoxal phosphate is involved in a large number of reactions related to amino acid metabolism. The vitamin is widely distributed in foods and is fairly heat stable, although in some vegetables, much of it may be present as the unavailable glycosides (i.e. linked to sugars). Deficiency is rare, although in the 1950s in the UK, an infant milk preparation which had undergone severe heating during manufacturing was found to have lost much of its pyridoxine, and infants fed almost entirely on the milk developed a number of metabolic abnormalities as a result of this deficiency (DoH, 1991b).

Vitamin B12

Vitamin B12 provides the only known requirement for cobalt in human metabolism, and comprises a corrinoid ring around an atom of cobalt. It is required for folate recycling and is closely associated with folate in a number of metabolic reactions. It is widely distributed in animal products, some algae (e.g. seaweeds) and bacteria, but does not occur in green plants. Fermented soy products such as miso and spirulina algae do not contain an active form of the vitamin, nor does yeast. Therefore, strict vegans may obtain virtually no vitamin B12 from food, unless consuming fortified products.

Folate

Folic acid (pteroyl glutamic acid) is the parent molecule of a large number of derivatives known as the folates. Rich food sources of folate include liver, yeast extract and green leafy vegetables, where the main form of the folate is as a conjugate with a string of glutamate molecules. The folate has to be hydrolysed to the monoglutamate form before it can be absorbed through the intestine, but the polyglutamates are the forms that are active within cells.

The main focus of health advisors regarding folic acid recently has been in ensuring that pregnant women consume enough to help ensure proper development of the baby's nervous system. A shortage of folate has been linked with the occurrence of neural tube defects, and a daily supplement of 400 µg per day, during the first 12 weeks of pregnancy, has been recommended by medical experts.

Pantothenic acid

This is a vitamin that is involved in the metabolism of all the macronutrients to produce energy. It is widely distributed in foods - meat, fish, poultry, whole-grain

Box 21 - Fortification of flour with folic acid

In the United States, the fortification of grain products with folic acid at 140ug/100g has been compulsory for several years and a 19% reduction in the level of neural tube defects as a consequence of this has been claimed (Honein *et al*, 2001). In the UK, the Committee on Medical Aspects of Food and Nutrition Policy reviewed the whole situation regarding the advice given to pregnant women and those considering becoming pregnant. In addition to recommending that current advice to these women to take 400µg folic acid per day as a medicinal or food supplement should continue, it also concluded that "universal folic acid fortification of flour at 240µg/100g in food products as consumed would have a significant effect in preventing neural tube defect-affected conceptions and births without resulting in unacceptably high intakes in any group of the population (DoH, 2000). The Committee acknowledged the problems that might arise, such as the masking of some of the symptoms of vitamin B12 deficiency, and it is this factor that Wharton and Booth (2001) emphasise in their urging of caution over fortification regimes.

cereals and legumes are particularly good sources - and, although it is fairly heat-labile, there is no evidence of any deficiency in the vitamin occurring in humans.

Biotin

Biotin is another vitamin involved in energy-providing metabolic systems. It is widely distributed in foods, and is only required in small amounts. Deficiency symptoms are rare, although they can be induced by eating large amounts of raw egg white (equivalent to 20-30 eggs per day). Egg white contains a glycoprotein, avidin, which binds very strongly to biotin, preventing its absorption.

Intestinal bacteria also produce biotin; it seems that this can be absorbed, but the extent of absorption is not known.

2.8 Other beneficial components

In addition to the nutrients listed above, food contains a vast complex of chemicals, many of which may have some nutritional or pharmaceutical value in certain circumstances. Given the variety and number of these chemicals, it is not surprising that some of them have been demonstrated to have effects in the body that can be shown to be beneficial. Many medical drugs are derived from plant sources in particular, and many plant foods are likely to contain chemicals that have some comparable effects. It is also likely that 'deficiency' in one of them would normally not be noticed, because of the myriad of other components having a similar beneficial effect. In some cases, because of the wide distribution of a component, a deficiency would be unlikely to occur, without more severe under-nutrition or mal-nutrition becoming evident first. Often the presence of one component will have a 'sparing' effect on another. This is seen with vitamins C and E and selenium, which have synergistic effects, although all are required individually for specific functions (Langseth, 1995). Many of these chemicals may have other significance in the food - they may be flavour compounds, or give it colour, or act as natural antioxidants or preservatives. It would also be misleading to suggest that, because a certain food has a chemical that has been shown to be beneficial, significantly increased intake of that food will have a comparable beneficial effect. As has been emphasised previously,

Box 22 - Breakfast cereal fortification with vitamins and iron

Cereal grains are good sources of some B vitamins and minerals, but the processes involved in the production of breakfast cereals are quite severe, and especially involve significant amounts of heat input. As a result, there is significant depletion of some of these nutrients. Replenishing these lost vitamins and fortifying with additional nutrients has been carried out for many years. Although vitamins are the main target for fortification, minerals can also be added, iron being the prime example. Although breakfast cereals are not unique in being fortified, they are one of the most significant food groups to be fortified, because of the significant contribution they make to total energy and protein intake, and the frequent and widespread use of them by virtually all segments of the population (Fast and Caldwell, 1990). It is also relatively easy and inexpensive to add vitamins to this type of product.

Breakfast cereals fall into two main categories - ready-to-eat (e.g. corn flakes) and products to be prepared (e.g. porridge oats). In the UK, the level of vitamins and minerals added to ready-to-eat cereals is fairly standard. Figures for typical toasted puffed rice products available in the UK are given below. All fortification levels are designed to give at least 100% of the RDA in 100g of product (as calculated from levels set down in legislation).

Nutrient	Level (per 100g)
Thiamin	1.4mg
Riboflavin	1.6mg
Niacin	18mg
Vitamin B6	2.0mg
Vitamin B12	1μg
Folate	200μg
Vitamin D	5.0μg
Pantothenate	6.0mg
Iron	14mg

In ready-to-eat products like cornflakes, vitamins such as niacin, riboflavin, vitamin B6 and vitamin E can be added to the cereal mix before processing, being relatively stable, but thiamin and vitamins A, D and C need to be added to the almost final product as a spray coating. This itself is not always straightforward as some vitamins are prone to oxidation, and spray systems maximise their exposure to oxygen. The incorporation of other antioxidants (e.g. butylated hydroxyanisole - BHA) as well as oxygen barriers (such as sucrose) can alleviate this problem (Fast and Caldwell, 1990). In some cases all of the nutrients may be added as a spray, thus simplifying the process.

food comprises a very complex array of chemicals, and there may be less desirable components in the food that we would not want to consume in greater amounts.

The variety of chemicals that comprise food is so vast that it is not possible to begin to analyse the role that each one might have from a nutritional point of view. Sadler *et al* (1999) deal with many of these macrosystems in great depth. However, there are some constituents that are currently being highlighted in foods, and some of them have been included in specially formulated products. Some of these chemicals occur in a wide range of foodstuffs, or can be synthesized by the body, while others are specific to certain foods or are added to foods as supplements. The term 'functional food' is often applied to this type of product. Brief consideration of a few of these topical food chemicals is given below, although it is beyond the scope of this book to comment in depth on the validity of some of the claims made about them.

Creatine

Creatine in its phosphorylated form (phosphocreatine or creatine phosphate) acts as a reservoir of phosphate for the production of ATP (adenosine triphosphate) from ADP (adenosine diphosphate). ATP is responsible for all energy-requiring activities in cells (it is hydrolysed to ADP and phosphate to release that energy). In physical exercise, the amount of ATP in muscles will only sustain contractile activity for less than a second. Therefore, the ADP that results must be immediately recycled (i.e. ATP must be reformed from ADP and a phosphate source) to sustain activity. Vertebrate muscle contains a reservoir of creatine phosphate to assist in this - it is the major source of phosphate in the first 4 seconds of a 100-meter sprint (Stryker, 1988). It is therefore being included in formulations that are claimed to be suitable for 'explosive' sports activities such as weight lifting.

Taurine

Taurine is a non-protein amino acid. It was originally only thought to be involved in the conjugation of bile acids to form taurocholic acid, but is now thought to have a role in the central nervous system. It has been shown to have positive functional benefits when added to infant formula milk (breast milk contains high concentrations) (Sadler *et al.*, 1999). It is not a dietary essential, as it is normally adequately synthesised from cysteine, and although there are very few plant sources of taurine, plasma levels in strict vegetarians are only slightly lower than those in carnivores.

Carnitine

Carnitine is synthesised in the body from the amino acids lysine and methionine. Although preformed carnitine is mainly associated with meat products (from where it gets its name), strict vegetarians who take in only about 10% of the levels consumed by carnivores still have normal body levels (Sadler *et al*, 1999). Carnitine is involved in the transfer of fatty acids into the mitochondria where they are oxidised, and it has been promoted as a nutraceutical beneficial to metabolic activity in a number of ways since the 1950s.

Phytoestrogens

Phytoestrogens are found in a number of edible plants. The two main chemical groups are the lignans and the isoflavonoids. The isoflavonoids occur mainly in legumes, especially in soya beans and soya products. The lignans occur mainly in seeds and grains. There appears to be an inverse link between some cancers and diets rich in phytoestrogens, with the incidence of breast and prostate cancer being much lower in Far Eastern countries, where there is an abundance of dietary phytoestrogens, than in Western countries.

Prebiotics

These are non-digestible oligosaccharides that support the growth of bifidobacteria in the colon, so changing and possibly improving colonic microflora. In contrast, probiotics are preparations of microorganisms added to food for the same end result.

Antioxidants

Langseth (1995) presents a concise monograph on the role of oxidants in various diseases. Highly reactive oxygen species such as hydrogen peroxide, singlet oxygen, ozone and various free radicals are formed during normal metabolic processes.

These are usually very short lived, and the body has various mechanisms to mop them up. Foods, especially plant foods, contain many chemicals with antioxidant activity. As well as vitamins C and E, mentioned earlier, several carotenoids have antioxidant activity. Brief details of their occurrence and that of some other antioxidants are given in Table 5.

Table 5 - Antioxidants obtained from foods

Antioxidant	Foods
Beta-carotene	many coloured fruits and vegetables
Alpha-carotene	carrots
Lycopene	tomatoes
Lutein/Zeaxanthin	broccoli, dark green leafy vegetables
Phenolic acids	soya beans, red wine, coffee
Isoflavones	soya beans
Catechins	tea
Polyphenols	tea, olives
Quercetin/kaempferol	onions
Flavonoids	fruits

3. NON–NUTRIENT CHEMICALS IN FOODS

All food is of animal, plant or microbial origin. As such, the original living organism manufactures and uses many chemicals that will not be required by another organism that uses it as a food. Most of these are completely innocuous at the levels consumed, having no role in or effect on human nutrition. Some have specific properties that maintain the structure and composition of the food (e.g. pectin in fruit). Others, as described above, are actually beneficial to human health. Others that may cause health problems, we have learned to deal with. A common example of this is the lectins in kidney beans; these have to be rendered harmless by soaking and vigorously boiling the beans before eating them. A more extreme example is puffer fish, a delicacy in Japan, which must be prepared in a precise fashion as it contains a lethal toxin. Others we just live with – like the dozens of compounds in onions that make us cry. This chapter describes some examples of these groups of compounds and looks at how food manufacturing is affected by them.

3.1 Flavour and taste compounds

The majority of flavour chemicals have no role in nutrition at all (although some nutrients such as sugars do contribute to flavour), but they do actually help to persuade us to eat the food that we need to eat, and are therefore quite important! Of course, at times, they may persuade us to eat things that perhaps we could do without (at least in the volumes that we consume). Processing of food will actually change the nature and level of flavour compounds in a food – fresh food will have a different flavour to the canned equivalent, and cured meat will be different to the uncured product. As well as occurring in the raw food, flavour compounds are regularly added to compound and processed foods. It is quite common for industrially produced flavour mixtures to have up to 50 individual chemicals in different proportions to provide the exact flavour required, and around 2000 individual flavour compounds are used in various combinations by the food industry. This gives an interesting comparison with the variety of compounds that are often found in raw materials. Individual raw fruits, for example, often contain two or three hundred compounds that contribute to a greater or lesser degree to their overall flavour. However, a very similar flavour can probably be reproduced artificially with just a small percentage of these.

Although classed separately from other additives in legislation, flavours are the most numerous and widely used group of chemicals added to food. They may be synthetically produced via industrial processes or may be of biological origin (i.e. extracted from an animal or plant – often termed 'natural'). The synthetically produced chemical may, in fact, be exactly the same as one that occurs naturally – i.e. 'nature-identical'. There has been much hype over recent years implying that 'natural' flavour chemicals (or other additives) may in some way be intrinsically less harmful than chemicals produced in a laboratory. There is no biological reason for this and it is important to judge the potential toxicity of each chemical on a case-by-case basis, irrespective of its origin. What is probably of more importance is the amount that occurs in a food, and the amount of that food that might reasonably be consumed. When using a naturally occurring flavour compound in a different setting, this has to be taken into account. Recently, the Scientific Committee for Food of the European Union published its opinions on two related compounds: estragole (1-allyl-4-methoxybenzene) and methyleugenol (4-allyl-1,2-dimethoxybenzene) (SCF, 2001). Both occur in a variety of plants: both are found in tarragon, basil and fennel varieties, and methyleugenol is also present in nutmeg, pimento and lemongrass. Both are also added as flavourings to products as diverse as jellies and ice cream, chewing gum and baked goods. The SCF's opinion was that both had been demonstrated to be genotoxic and carcinogenic at certain levels, and that a general reduction in intake via flavourings should be considered through restrictions on the levels used (although that does not mean that levels currently used are in any way hazardous). There is no suggestion that levels consumed in herbs and spices pose a problem. Most, if not all, food that we consume will contain flavour or other compounds which, if consumed in excess, might lead to health problems. Only a few are likely to cause problems at lower levels, and examples of these are discussed in a following section.

Processing may change the flavour profile of a food by a number of mechanisms: heating will alter or in some cases completely destroy individual flavour compounds, or cause them to be simply 'boiled away'. It may also create new ones from chemicals that themselves have little or no flavour. Destruction of cell walls through heating or freezing may mean that enzymes come into contact with substrates that are normally kept separate; the ensuing chemical reactions result in comparable flavour effects. The overall result is a change in the flavour profile (due to changes in the levels of the individual chemicals that make up that profile). This change may be desirable or not.

Historically, food processing has involved making raw food more palatable, sometimes producing an edible product from inedible starting materials. One factor in this is to make the food taste better. However, sometimes there are other over-riding constraints. The canning industry was developed as a way of preserving fish, meat, fruit and vegetables to make them available to a wider number of people for a greater period of time through the year. The degree of heat processing required in canning and other thermal processing treatments to produce a microbiologically safe product will usually significantly alter the flavour. In many cases this change will be, to a certain extent, undesirable. Therefore, the food industry is continually looking for ways to reduce the amount of thermal processing given without compromising food safety.

There are two general approaches to doing this. The first is to understand better how much processing is required to make a product microbiologically safe. In the absence of this knowledge, large safety margins have to be built in, resulting in over-processing and a greater than necessary loss of quality, including flavour. The second way is to combine heat processing with other forms of processing which have a lesser effect on flavour and overall quality, or which make the heat processing itself intrinsically more effective. Examples are the use of ultrasound, extreme high pressure (EHP), and ultraviolet and high-intensity light. In some cases, these new processing techniques may be able to be used as a sole preservation technique (see box).

Box 23 - Traditional ways of altering flavour profile

Many traditional food and drink products and processes are based on the very precise, but completely empirical, altering of the flavour profile of ingredients, e.g. brewing, spirit distilling, wine making, and cheese making. Each of these involves a basic process, minor and subtle alteration in which give rise to thousands of individual products, each with its own distinctive flavour qualities.

In cheese making, changes in starter culture, milk pretreatment and ripening stages can all have significant impacts on the end product. The amount of a simple ingredient like salt can also have wide-ranging effects. In a study of reduced-fat Cheddar cheeses, Mistry and Kasperson (1998) found that increasing salt levels in the water phase from 2.7 to 4.5% reduced bitterness intensity. In a similar study by Banks *et al* (1993), cheddary flavour was found to increase with increasing salt levels from 1.17 to 1.99%. The bitterness defect at low salt concentrations is probably due to the inhibition of beta-casein hydrolysis.

Efforts have been made with some products to look at all the variables to optimise flavour and quality characteristics. In such a study of Serra cheese (Portugal's principal traditional cheese manufactured from raw ewes' milk), Macedo and Malcata (1997) found that the addition of 0.05g salt per sq. cm of fresh cheese surface resulted in the highest sensory scores, as well as the lowest counts of viable enterococcal and coliform bacteria.

The development of soy sauce is an interesting example of how flavour profiles have been developed empirically without fully understanding the chemical changes underpinning them. Soy sauce (shoyu) is usually produced from a soybean/wheat kernel mixture by yeast fermentation. Koikuchi is by far the most popular and is the type regularly used in the UK. Salt levels of 17-18% have traditionally been used in shoyu fermentation for optimum development of the flavour profile. However, it was not until 1996 that Sasaki showed that this was indeed the optimum level for the production of many of the individual chemicals that contribute to the overall flavour of the sauce.

Box 24 - Extremely high pressure - a novel method for improving flavour

High-pressure was first proposed as a method for processing food in the late 19th century, but it was not until the 1970s that companies in Japan started to exploit the technology commercially to produce products of good microbiological quality that retained more of the flavour attributes of the fresh product.

Orchard House Foods has recently introduced 'Extremely High Pressure' (EHP) treated orange juice to the UK. They claim that a 'low temperature, high pressure system' that kills bacteria naturally present in fresh foods, whilst significantly extending shelf life, is ideally suited to fresh orange juice. Because the EHP treated juice is not heated, as it is with pasteurisation, the flavour profile in the orange juice is not affected in the same way.

In taste tests conducted by Orchard House to compare fresh, EHP and pasteurised orange juice, EHP juice emerged virtually identical to fresh, whilst pasteurised juice was felt to taste more processed, less sweet and less tangy and juicy. 88% of consumers stated that EHP treated juice tasted 'fresh' compared with just 56% who thought the same of pasteurised juice. From a visual perspective, nine out of ten consumers also thought that EHP juice looked natural and appetising, compared with 73% who felt this was true of pasteurised juice.

The company claims that EHP treated orange juice retains all the natural characteristics of fresh orange juice whilst extending shelf life from its current nine days to up to 21 days.

Orchard House is the first UK company to install this particular piece of technology, but the system has already proved successful in the US and mainland Europe where it has been used to extend the shelf life of short shelf-life products such as guacamole and fresh apple juice. It has also been used successfully to process oysters without loss of flavour (Orchard House Foods, 2001).

As well as modifying processes to reduce flavour loss, or to create the desired overall flavour, many processes incorporate steps or additional ingredients to prevent the development of off-flavours or other undesirable flavour changes. Preventing the development of rancidity is one example. Rancidity is the sensory perception of specific changes in the nature of the lipids in a food. There are several types of rancidity, the two main types being oxidative rancidity and hydrolytic rancidity. In hydrolytic rancidity, the triglycerides are broken down to release free fatty acids. In oxidative rancidity, the fatty acid chains themselves are oxidized (typically a double bond in the chain is broken and oxygen is incorporated). Both can result in the perception of an off-flavour. In minor cases, the off-flavour is, or can be, masked (e.g. by the incorporation of highly flavoured ingredients such as spices into the final product) and the problem is circumvented, but this is not always possible or desirable, and preventative measures are required. Cooking oils, for example, are prone to various chemical changes, some of which result in rancidity (they also become less efficient at frying). Ways of delaying this include the addition of permitted antioxidants (such as butylated hydroxyanisole, butylated hydroxytoluene and gallate), not overheating the oil, keeping it dry (i.e. drying products before frying where possible) and removing debris from the oil that would otherwise act as a catalyst for some of the chemical reactions involved.

Fatty fish such as herring and mackerel contain a high proportion of unsaturated fatty acids and are prone to oxidative rancidity development. The rate of oxidation is dependent on temperature, but still occurs in the frozen fish. Antioxidants have proved to only have a limited effect, and preventing the access of oxygen to the fish, in combination with low temperature, is the best way to slow the development of rancidity. Glazing, vacuum packaging and coating with batter are three possible routes; preventing dehydration also helps (Ranken *et al*, 1997).

The overall flavour of all food, both animal and plant, will vary depending on the stage in its life cycle at which it was harvested. Beef is different to veal, although the same individual animal could be used to provide either, as could lamb and mutton; and the pods of mange-tout peas are considered edible, whereas pods of main crop peas are not usually eaten. The flavour profile will also begin to change immediately the plant or animal is 'harvested'. Some of these changes are beneficial (e.g. meat and poultry products may be 'hung' after slaughter to develop flavour; fruits continue to ripen after being picked - unripe bananas are noticeably different to ripe fruit), whereas others are not desirable. These deleterious changes may be brought about mainly by external factors, such as fat oxidation in oily fish as described above, or they may result from enzymic action remaining within the food. Some vegetables are particularly prone to undesirable changes over quite a short period of time and need to be blanched before being further processed (see box).

Box 25 - Prevention of off-flavour formation in frozen vegetables

Producing frozen vegetables of acceptable flavour quality is more than just a case of picking or digging them from the field, washing or peeling and storing them at an adequately low temperature. The harvesting process itself can result in some bruising, which may promote undesirable enzyme activity. Cell walls and membranes become damaged, allowing enzymes to gain access to substrates from which they would normally be separated. Even without this damage, enzyme-mediated off-flavour development will begin to occur. Although frozen storage slows down this activity very significantly, the length of time over which frozen food is kept means that changes would be detectable within the normal shelf-life. Off-flavours in peas can be detected within 3-4 hours of harvest. The sugar in peas also begins to be converted to starch quite rapidly after harvesting, leading to a less sweet product if this is not prevented.

To prevent the deterioration in flavour, most vegetables are blanched (or scalded) before being frozen. This involves holding the vegetable in water at 95˚C or steam for 2-3 minutes. The exact processing conditions will vary from one vegetable to another, and between varieties. The main deleterious enzymes that are destroyed are peroxidases, catalases and lipoxygenases. Peroxidase is the most heat-resistant enzyme found in vegetable systems, and the effectiveness of the blanching process is usually measured by determining if peroxidase activity has been eliminated. Heat treatment is also likely to have some detrimental effects on product quality, such as vitamin loss and leaching of minerals (although it can prevent enzymic breakdown of some vitamins), and so the process has to be closely controlled and adapted to the needs of the particular vegetable. Care needs to be taken with the physical integrity of delicate products such as cauliflower florets. Steam blanching, although technically more difficult than water blanching, does minimise some of the nutrient losses, and can give the product a longer shelf-life.

Blanching procedures are also used to pre-treat vegetables before canning or dehydration preservation.

3.2 Colours

The colour of a food is one of the most important attributes in determining whether or not we decide to eat it - subsequent flavour or texture attributes may make us change our mind, but colour is of great importance in making the initial decision, and the industry puts much effort into satisfying consumer demand for food of the 'right' colour.

The colour of a food is determined by the combination of chemicals present in the food. As with flavour, colour can change with processing, and in some instances processing procedures are modified to either reduce or modify that colour change, or indeed enhance it if the final colour is deemed to be more desirable that the initial colour. In other cases, colours are added to products before or after processing, sometimes to mimic the original colour of the raw material.

There are a variety of chemicals that have a role in the overall colour of raw foods. In fruits and vegetables, the primary colouring chemicals are chlorophyll, carotenoids and anthocyanins. In raw red meats, the colour is due to haemoglobin. During storage or processing, the existing colours in these raw materials may change. The most widespread colour change seen in raw foods is that seen during the ripening of fruit: tomatoes turn from green to red; peppers may stay green or turn red, yellow or orange, depending on the variety; the skins of bananas also change from green to yellow, as one colouring material is lost from the plant and others are synthesised. These changes indicate, in many cases, that the fruit is ready to eat, thus providing a method for the parent plant of distributing its seeds.

The changes that occur during the storage of raw foods are generally subtle and in some cases desirable, but it still takes a concerted effort from grower or producer through to the retailer to get the products onto the shelf at the right time. The ripening of fruits can be controlled by storage in controlled and/or modified gas atmospheres, so that they are presented for sale at the optimum time. The colour of raw meat is also important in its appeal to the consumer; the bright red colour of red meats is due to haemoglobin being in its oxygenated form, and this can be optimised by packing cuts of meat under a modified atmosphere containing high levels of oxygen.

The colour of egg yolk is significantly affected by the diet of the hen that laid it, and feed can be supplemented with additives to produce a deep yellow colour in the eggs, which has traditionally been preferred. Recently, there has been a move away from the use of these additives, and pale-yolked eggs have been deliberately marketed as being a result of the removal of these feed additives. The incorporation of additives into feed has also been practised for many years in fish farming, with the carotenoid canthaxanthin being used to increase the pink colour in salmon. There has been some concern about possible adverse health side-effects in man, with claims that very high levels of consumption of canthaxanthin result in the deposition of minute crystals in the eye. However, a recent review by Baker (2002) suggests consumption of even very high levels of pigmented salmon is unlikely to approach the levels required to have this effect, and that the effect is both non-damaging and reversible.

Processing can have a more significant, sometimes deleterious effect, on food colour and food manufacturers have to take steps to either minimise these changes or compensate for them. The processing (canning) of garden peas destroys the green colour and without the addition of colouring, the product would appear grey and rather unappealing. Only three colours are permitted to be added to this type of product in the EU - tartrazine, brilliant blue FCF and Green S - specifically to compensate for the loss of natural colour.

A comparable situation exists with the canning of strawberries, which lose their red colour during processing and become brown. Erythrosine was traditionally added in the UK to compensate for this, but this is now not permitted in the EU. Further information on the use of colour additives in food is given in the next section.

Processing can also be optimised to produce desired colour changes in food. The browning of bread crust and darkening of coffee during roasting, for example, are due largely to the production of a series of chemicals known as Maillard reaction products, which occurs through reactions of sugars with amino acids and may be controlled by the process conditions.

Other browning reactions are not so desirable, specifically the darkening of the cut surface of fruit and vegetables, due to the action of the enzyme polyphenol oxidase on phenols (see Box 26).

Box 26 - Preventing enzymic browning

The cells of fruits and vegetables naturally contain the enzyme polyphenol oxidase (PPO) and chemicals called monophenols. In healthy tissue the PPO and the monophenols are held in separate compartments within the cell. When the fruit or vegetable is cut (e.g. sliced, peeled, shredded, trimmed, cored) the PPO and monophenols get mixed, at which point the PPO catalyses the conversion of the monophenols to quinones. These quinones react with oxygen in the air, in a chemical oxidation reaction, to form large complex polyphenol molecules - the brown pigment that causes the discoloration.

Various approaches have been developed to retard or prevent this browning and so prolong the shelf-life of prepared fruit and vegetable products. These target one or more steps in the series of events that lead to the formation of the brown polyphenolic pigments and examples include:

- Sulphite dips - Sulphite prevents enzymic browning in two ways: it inhibits the action of polyphenoloxidase, slowing down the formation of quinones from monophenols, and it reacts with quinones to form stable intermediates that do not further react to form polyphenols.

- Non-sulphite dips such as ascorbate can also prevent browning by interfering with the oxidation of quinones. In this case, the ascorbate reduces the quinones back to monophenols before they form the brown polyphenolic pigments, reversing the effects of PPO. Unfortunately, once the ascorbate is consumed in this reaction, browning will resume.

- Use of modified atmosphere packaging such as high oxygen atmospheres (80-95% oxygen with the remainder as nitrogen) which inhibit enzymic browning of fresh cut produce. It has been hypothesised that the effect is a result either of oxygen (i.e. substrate) mediated inhibition of PPO or quinone (i.e. product) feedback inhibition of PPO (Day 2001).

- Edible coatings that contain natural anti-oxidants such as ascorbate have been developed for use with peeled and sliced fruit prone to browning - most notably apple - providing an alternative to standard liquid dips (CCFRA 1999). Some of these might also form a layer impermeable to oxygen to reduce the rate of chemically mediated oxidation of any quinones formed.

continued....

◆ Genetic modification has been used to block the formation of the polyphenol oxidase enzyme (Bacchem *et al*, 1994). This was done using so-called 'anti-sense technology' to 'turn off' the gene for PPO. Without this, the tissue was unable to convert the monophenols to quinones, so that the whole chain of reactions was blocked and enzymic browning was prevented. This was first applied successfully to potatoes and demonstrated on a pilot scale.

References:

Bacchem, C.W.B., Speckmann, G.J., van der Linde, P.C.G., Verheggen, F.T.M., Hunt, M.D., Steffens, J.C. and Zabeau, M. (1994) Antisense expression of polyphenol oxidase genes inhibits enzymatic browning in potato tubers. Bio/Technology 12 1101-1105.

Baker, R.T.M. (2002) Canthaxanthin in aquafeed applications: is there any risk? Trends in Food Science and Technology 12 (7): 240-243.

CCFRA (1999) New Technologies Bulletin No. 19

Day, B. (2001) Fresh prepared produce: GMP for high oxygen MAP and non-sulphite dipping. CCFRA Guideline No. 31 - Campden & Chorleywood Food Research Association.

3.3 Additives

In addition to the chemicals that naturally make up the raw materials that we eat, other chemicals are often added during the processing or preservation of food – for a variety of reasons. There are many classes of additives that are routinely used in processed food production (see Table 6). As with flavours, these may be synthetic, 'natural' or nature identical. These chemicals fulfil a variety of roles in foods and their use is very closely controlled. All additives in use in the European Community are evaluated from a toxicological point of view; they are only permitted in foods where there is a proven technical need, and then only at levels that would not be expected to give rise to health concerns. Far from being 'bad', the E number system is proof that the compound has undergone these rigorous evaluations. In the USA, a similar and equally stringent approval system is operated. The vast majority of additives permitted for use in the EU have been used over a long period of time and have a proven safety record in actual use, as well as in laboratory trials.

Food chemical composition

In the UK, three major pieces of legislation cover the use of additives in foods. Derived from EC Directives these cover the addition of colours, artificial sweeteners, and 'miscellaneous' additives to food intended for sale to the public. The 'miscellaneous' category actually includes such things as preservatives, antioxidants, emulsifiers, stabilisers, acidulants, humectants, anti-caking agents, anti-foaming agents, flavour enhancers and thickeners amongst others. In all three pieces of legislation, there are positive lists of what is permitted to be added to foods, and in many cases there are restrictions on the amount that can be added and the foods to which they can be added. In addition, there is legislation covering the presence of associated materials such as extraction solvents, flavourings and mineral hydrocarbons.

It is beyond the scope of this book to discuss in detail the role of each of the groups of permitted additives and any associated health issues, but a few examples are given below.

Table 6 - Additive categories routinely used in food production

Acid	Foaming agent
Acidity Regulator	Gelling agent
Anti-caking agent	Glazing agent
Anti-foaming agent	Humectant
Antioxidant	Packaging gas
Bulking agent	Preservative
Colour	Raising agent
Emulsifier	Sequestrant
Emulsifying salt	Stabiliser
Firming agent	Sweetener
Flavour enhancer	Thickener

Preservatives are added to formulated food to prevent or retard microbiological growth, either from microorganisms that may be inherently present in the food, or those which may gain access to the food during its production and storage, including after sale. The consumer-driven move to products with less sugar or salt

60

has meant that some products are less microbiologically stable; the addition of preservatives is one way to offset this and maintain a relatively long shelf life. Jams and marmalades are typical examples of reduced-sugar products. However, there is also a consumer-led desire for a reduction in preservative usage. In some cases this will mean that the product will have a reduced shelf-life, or will need to be refrigerated or both. The most widely used preservatives are sulphur dioxide and the bisulphites, sorbic acid and its sodium and potassium salts, benzoic acid and its salts, and the parahydroxybenzoates. Additionally, nitrates and nitrites are used in cured meat products, but the few other preservatives permitted in the European Union have very limited and specialised uses.

The primary function of antioxidants is in the prevention of rancidity in fatty products. Again, the number permitted in the European Union is limited, with butylated hydroxyanisole, butylated hydroxytoluene and the gallates being the major ones. Ascorbic acid (vitamin C) is also a permitted antioxidant. It is interesting to note that when it is used as a vitamin, it has to be declared as such on the label; similarly, when it is used as an antioxidant (i.e. its purpose is to have an effect on the food, not on the consumer!), it must be declared as the chemical name or 'E' number.

There are many colourings permitted for use in food. Many of these are derived from plants, one (cochineal) is derived from an animal source, and some are entirely synthetic. There has been much written about the acute effects of some of the synthetic colours, particularly the azo or coal tar dyes (azo refers to the presence of a nitrogen-nitrogen double bond in the molecule), with regard to their inducing hyperactivity in children, although the nature of the link has not been conclusively identified. Current legislation generally restricts their use to certain products and certain maximum amounts. However, some 'natural' colours also have to be controlled. As with flavourings, the 'natural' colours can occur in foods associated with their natural origin, e.g. curcumin in turmeric, and there is no limit to the amount of turmeric that can be added to a product. However, the level of curcumin itself that can be added to other foods is closely controlled, as is the list of foods to which it can be added.

Many other additives have multiple processing functions, depending on the food type. The many permitted phosphates, for example, are used as buffering agents, sequestrants, emulsifiers, stabilisers, texturisers and raising agents.

Finally, other additive-type chemicals appear in food because they were used in the processing of the food. These processing aids have no function in the final product, and are merely 'left over' after processing because it is not feasible to remove them (e.g. lubricant on a sausage casing that facilitated the extrusion of the sausage). As with all chemicals added to food, general food law requires the food manufacturer or processor to ensure that the levels occurring in the food are neither injurious to health nor adversely affect its quality to an unacceptable degree.

Box 27 - Additive usage in food products

There has been considerable consumer pressure for a reduction in the volume of additives used in formulated food products in recent years. Much of this has focussed on 'artificial' colours, sweeteners and preservatives, although the meaning of the word artificial in this context is not always clear. Nevertheless, the food industry has responded by developing new versions of standard products with reduced levels or numbers of certain additives, and has remarketed others that are already low in these chemicals. Over 50 new products launched in 1999 and 2000 contained the legend 'no artificial colour', which in this case means no added colouring chemical or extract (rather than just no synthetic colour) that would need to be labelled as an additive. Thus, although turmeric could be present as an ingredient in a curry paste product labelled as 'no artificial colour', giving it a yellow colour, curcumin (the yellow colour itself) could not be added in isolation if the product is to carry the declaration.

There have been some significant changes in product formulation as a result of consumer demand. One example is the colour content in canned soft drinks. Given the fact that these are often drunk straight from the can, the use of synthetic bright yellow and green colourings such as tartrazine and sunset yellow might be seen as an unnecessary expense and a negative marketing point. Many of these drinks are now formulated to contain only beta-carotene or other similar colours of biological origin that are more acceptable to the consumer.

Reformulation of products with reduced artificial colours or sweeteners will merely produce less colourful and less sweet products, unless either naturally coloured and/or sweet ingredients are used in their place. Reducing the level of preservative used in formulations will require either a different storage regime, a shorter storage time (reduced shelf life) or significant reformulation or packaging. In some cases this may require specific instructions to the consumer, perhaps to store refrigerated after opening, when they may be used to similar products being stable at room temperature.

Box 28 - Getting a novel ingredient or additive approved

The legislative control over additives and ingredients varies throughout the world. In the UK and EU, any ingredient can be used as a food or in a food, providing that it has a history of safe food use in the EU. However, any novel ingredient has to go through an approval process to assess its efficacy and safety. In 1997, the EU harmonised novel foods legislation throughout the Community (European Parliament, 1997). This was enacted in the UK as the Novel Foods and Novel Food Ingredients Regulations.

In simple terms, the person responsible for placing the novel food or ingredient on the Community market must submit a request to the Member States in which the product is to be marketed for the first time and must also forward a copy of the request to the European Commission. In the UK, requests are assessed by the Advisory Committee on Novel Foods and Processes (ACNFP). An initial assessment is carried out within 90 days, and this is then forwarded to the Commission who pass it on the other Member States. They have a further 60 days to comment. If the original assessment is favourable and no objections are subsequently raised, then the novel food or ingredient can be marketed. If there are objections, the application is referred to the EC Standing Committee for final agreement. (This body also rules on whether a food or ingredient is novel, in cases of doubt or dispute.)

The type of data that might be needed to support an application would be evidence of safe use in other parts of the world, or a demonstration that the new food is broadly similar to an existing food along with supporting safety information. Many potential novel ingredients are complete biological entities - perhaps a tropical root, with a history of use as a food in another part of the world, or a new herbal extract with a similar background. Others are specific chemicals that count as ingredients because of their function, rather than as additives. One such example is trehalose, which has recently been approved by the European Commission. Trehalose is a naturally occurring, highly stable disaccharide. Although it is metabolised to glucose, its glycaemic effect is claimed to be much more even than that of other sugars (i.e. the rise and fall in blood glucose levels is less severe). It is less than half as sweet as sucrose and its suppliers believe it will provide a range of technical benefits in food products, such as: moderating sweetness in bakery products, processed fruits and confectionery products: replacing fat and controlling moisture migration in baked goods and confectionery products; and providing stable, resilient coatings in pan-coated products.

continued...

Similarly there are stringent controls on the use and approval of additives. To extend the use of a particular additive to a new food type or group of foods, or to obtain approval for use of a new additive requires 'safety' and 'need' aspects to be examined in detail. The need aspect will include commercial and cost implications (it may be cheaper to produce than an alternative additive performing a similar function), whilst safety aspects will take into account the volume of the additive that would be consumed as part of a normal diet. Minor amendments of permitted additive use occur from time to time, but completely new additions are rare.

3.4 Antinutrients and toxicants

It has been said that 99% of all toxins are naturally occurring, and also that all things are toxic at a high enough concentration. Certainly, many foods contain chemicals which, if consumed in excess, might lead to health problems. Many of the nutrients discussed in the previous chapter could be included in this group. Some foods contain certain chemicals at levels that could have significant adverse health implications even if eaten at 'normal' levels. Most of these we have dealt with by various traditional cooking and processing techniques over the centuries. Indeed, the advent of cooking drastically increased the amount of plant material that could be eaten, e.g. plants such as potatoes, which are high in starch (the human digestive system is not adapted to hydrolyse raw starch), and seeds and legumes, many of which contain toxins in the raw state, but become edible after cooking (Liener, 1980). Cooking also removes or inactivates many chemicals (e.g. protease inhibitors) that inhibit digestion or absorption of nutrients. However, some chemicals have arisen as problems associated with food processing techniques developed in the last 100 years or so, e.g. trans fatty acids resulting from chemical hydrogenation of unsaturated fats, or 3-monochloropropanediol derived from lipids during the chemical hydrolysis of proteins (to form HVP - hydroysed vegetable protein).

These problem chemicals may occur as regular constituents of the food in question (e.g. lectins in kidney beans), or at increased levels as responses of the foodstuff when alive to some sort of stress (e.g. glycoalkaloids in potatoes, an increased production of which can be stimulated when the tuber is exposed to light). Alternatively, they may be accumulated by the food during its lifetime, and be of no benefit to it (e.g. heavy metal accumulation in shellfish, or nitrate accumulation in leafy vegetables). There are also some instances of a processing regime potentially releasing a toxin from a non-toxic starting material (as occurs with cyanogenic glycosides in some canned stone fruits).

Toxins may have a direct adverse effect on health (e.g. by causing inflammation or blocking respiration), or they may act as anti-nutrients, e.g. by blocking specific nutrient absorption and so mimicking the effects of nutrient deficiency. Most of the potential toxins are innocuous in the amounts that we consume, and we have learnt not to eat an excess that might cause a problem. However, some require special attention by the industry and by the public in general. Some examples of these toxins are given below.

3.4.1 Lectins (Haemagglutinins)

Lectins occur in a wide variety of plants including beans of the *Phaseolus* genus (e.g. kidney beans - *P. vulgaris*, and lima beans - *P. lunatus),* broad beans (*Vicia faba*), castor beans, soya beans, lentils, peas, field beans (*Dolichos lablab*), peanuts, potatoes and cereals, as well as a range of non-food plants (Jaffe, 1973; 1980). In many cases the lectins have no or minimal toxic effect. Others are toxic to a greater or lesser extent, but in most cases normal cooking procedures eliminate this toxicity entirely, and consumption of moderate levels of most types of uncooked beans or peas will have no adverse effect. However, there are some specific exceptions.

There are many different types of lectins. They are glycoproteins and are examples of chemicals that have a variety of roles in the food while it is alive, but which have no significant positive nutritional role in our diet (they do not occur in large quantities and therefore do not act as a significant protein source). In plants it has been suggested that they act as antibodies to counteract soil bacteria, protect plants against fungal attack, attach to glycoprotein enzymes in organised multi-enzyme systems, and play a role in the development and differentiation of embryonic cells, and in the transportation and storage of sugars.

Their mode of action in the plant is probably related to their toxic activity in the diet. Their main physiological characteristic is their specific affinity for certain sugar molecules. Most animal cell membranes contain these molecules, and the lectins can be characterised by their ability to bind to and agglutinate red blood cells (hence the name haemagglutinins). However, their toxic activity derives from their ability to attach to the same sugar moieties on intestinal cell membranes. The nature of this toxicity can range from long-term impaired nutrient uptake to serious acute food poisoning symptoms. As stated above, most lectin sources do not cause a problem, and they are easily denatured by the heating processes occurring in normal cooking. However, there are specific exceptions, the most well known and significant of which (because of the way we consume them) is red kidney beans.

Raw kidney beans are significantly toxic and must be cooked adequately before consumption. Given the wide availability of dried, raw kidney beans in the UK marketplace, and adverse TV publicity in the summer of 1981, the then Department of Health and Social Security issued instructions to the food industry on the giving of general advice to purchasers of raw beans for their preparation. The form of wording suggested for the labelling of beans for sale was:

"After soaking overnight and throwing away the water, these beans should be boiled briskly for at least 10 minutes and then cooked until soft, otherwise they may cause stomach upsets. Never cook in a small casserole unless the beans have first been soaked and boiled in this way. Do not eat raw beans"

It was suggested that, because of possible confusion in the identification of beans by the consumer, that all dried beans should be labelled in this way. It was subsequently recommended that the reference to stomach upsets be removed in case this dissuaded people from purchasing dried beans. The reference to brisk boiling is the most important factor, as there is some evidence to suggest that gentle heating below boiling actually increases the lectin activity rather than destroying it.

3.4.2 Potato glycoalkaloids

Potato glycoalkaloids are a good example of naturally occurring toxins that can and have caused problems when consumed in large quantities, but which we have learnt to avoid without too much problem. However, at least 12 separate cases of food poisoning from potato consumption have been documented, involving around 2000 people and resulting in 30 deaths (Smith *et al*, 1996). Potato glycoalkaloids also have a bitter, burning taste, and can cause bitter taint problems. For these dual reasons, they present a natural hazard that must be controlled.

Potatoes contain two main glycoalkaloids: solanine and chaconine. Both are nitrogenous steroids (solanidine) linked to 3 sugar molecules, with chaconine being the more biologically active. Symptoms of acute poisoning can range from abdominal pain, vomiting and diarrhoea (similar to bacterial food poisoning) to confusion, fever, hallucination, paralysis, convulsions and occasionally death. There is an unofficial, but widely accepted safety limit of 200 mg glycoalkaloid/kg fresh potato. Levels of glycoalkaloids in modern varieties are usually well below this value, but they can exceed the limit under certain circumstances. The associated bitterness that accompanies these increases mean that the chances of ingesting a toxic dose are small unless the bitterness has been masked with other highly flavoured ingredients.

Thus, although the chances of someone eating potatoes with high levels of glycoalkaloids are small, the possibility does exist, and the food industry must take precautions to eliminate the risk as far as is possible. Glycoalkaloids can only be made by the living potato tissue, and therefore will be halted by cooking and any other process that kills the tissue. However, they are heat stable and therefore preformed toxin will remain after processing. Glycoalkaloid levels in potatoes are highest in the flowers and in the sprouts on the tubers. Within the mass of the tuber itself, they are concentrated in the outer 2mm, so that unpeeled potato products are a higher risk than flesh-only products. Levels vary from one variety to another, and are generally higher in early varieties than in main crop varieties. Smaller potatoes tend to have higher levels than large potatoes, largely as a consequence of the increased surface area/volume ratio.

Increased levels arise through various stress factors, such as pest and disease damage, drought, water-logging, and extremes of temperature. During post-harvest handling, bruising, abrasion and other types of mechanical damage can all cause increases in levels, as can peeling (although the act of peeling will remove much of the glycoalkaoid content unless the peel is added back into the product). Light can also induce glycoalkaoid formation. Light also induces chlorophyll formation, causing the potatoes to turn green on the surface. This has led to the partial myth that green potatoes are always poisonous. Light does not always induce glycoalkaloid formation, so not all green potatoes will pose a problem. Conversely, many non-green potatoes may contain high glycoalkaloid contents for other reasons. However, given the scope for confusion, it is best to assume that green potatoes will contain high levels of glycoalkaloids.

Key factors for the food industry in minimising the risk of using potatoes with elevated glycoalkaoid levels are:

- Know the history of the crop and use only good quality tubers
- Where possible use varieties known to have lower levels
- Avoid tubers which have been damaged by pests or disease, which have been stressed during growing or after harvesting, or which have been poorly stored or show signs of sprouting or greening
- Handle raw potatoes with care and avoid bruising or other physical damage
- Minimise the delay between peeling and processing potatoes
- Pay extra attention to the above if producing peel-on potato products

3.4.3 Oxalates

Oxalic acid and oxalates are widely distributed in plant foods, highest levels being found in spinach (0.3-1.2%), rhubarb (0.2-1.3%), tea (0.3-2.0%) and cocoa (0.5-0.9%). Many other foods, such as lettuce, celery, cabbage, cauliflower, carrots, potatoes, peas and beans contain up to about a tenth of these levels (Fassett, 1973). Although there is no question that the ingestion of sufficient oxalic acid as crystals or in solution can be fatal, there is considerable debate as to whether serious food poisoning from oxalate is usually due to food.

The eating of rhubarb leaves has been a well-known cause of illness for centuries. Rhubarb leaves contain high amounts of oxalate. However, the levels of oxalate in rhubarb stalks are sufficiently high that consumption of normal levels of rhubarb stalks will result in at least as much oxalate intake as from small to moderate amounts of leaves. Fassett (1973) disputes the conclusions and assumptions made about the causes of several food poisoning incidents which have been ascribed to oxalate and suggests that more attention should be given to anthraquinone glycosides, which are known to be present in rhubarb leaves (at 0.5-1.0%) and other related species of plant.

Whatever the toxic principle, consumer perception is, correctly, that rhubarb leaves are toxic and consumer complaints about small fragments of leaves in canned rhubarb are well-known. Although the levels generally encountered in food products are of no health significance at all, rhubarb leaves are unacceptable to the consumer and the manufacturer needs to make special effort to eliminate them. This basically means ensuring that all traces of leaves are removed from the stalks during the trimming stage before processing and that suitable separation systems are in operation to ensure that stray pieces of leaves do not become associated with the stalks further down the processing line.

As well as an acute toxic effect, the chronic effect of oxalate has been hypothesised. Because of its structure, oxalate can bind calcium and interfere with its metabolism, including its absorption, but there is little evidence that levels consumed could be high enough to have a significant enough effect on calcium levels in the body that would result in a deficiency syndrome.

3.4.4 Phytates

Phytic acid and the phytates have the ability to chelate (bind) divalent and trivalent metal ions such as calcium, magnesium, zinc and copper. As such, they have the potential to interfere with the absorption of these ions in the gut. Levels of phytate vary greatly with the stage of maturity of the plant and the portion of the plant that is consumed, but cereals, nuts and legumes have relatively high levels, and potatoes, sweet potatoes, artichokes, blackberries, strawberries and figs contain small to moderate amounts. In contrast, lettuce, onions, mushrooms, celery, spinach, bananas, pineapples, apples and citrus fruits are devoid of phytate (Oberleas, 1973).

Whether or not high levels of consumption of phytate-containing foods will result in mineral deficiency will depend on what else is being consumed. In areas of the world where cereal proteins are a major and predominant dietary factor, the associated phytate intake is a cause for concern. The major issues in the UK are more concerned with quoting mineral levels in food, when in fact the bioavailability of these minerals may be significantly less than the quantity in the foodstuff. More information on bioavailability is given in a preceding chapter.

Box 29 - Phytate as a source of phosphorus

As well as being a problem for mineral absorption in the mammalian diet, phytate itself is also a potentially good source of dietary phosphorus. Pig feed based on cereal grains and oil seed meals is high in phytate, and this phytate accounts for around 80% of the total phosphorus content of the feed. However, under normal circumstances, this phosphorus is unavailable, as the phytate cannot be metabolised. The two general strategies have been to augment the diet with mineral phosphorus, or to include as a supplement phytase, an enzyme that releases phosphorus from phytate. The former strategy results in phytate being excreted undigested, which has long been a problem for the pork industry because of the resultant environmental pollution from the high levels of phosphorus in the manure.

As a potential new strategy, researchers have now developed a transgenic pig that produces its own phytase in its saliva (Golovan *et al*, 2001). They report that the pigs seem to require almost no inorganic phosphate supplementataion for normal growth and that they excrete up to 75% less faecal phosphorus than non-transgenic pigs.

3.4.5 Enzyme inhibitors

Many constituents of plant foodstuffs are known to contain enzyme inhibitors. Liener and Kakade (1980) list nearly 100 that affect various proteases. Seeds are the most common part of the plant to contain protease inhibitors (they help the seed survive in the gastrointestinal tract as part of their dispersal strategy), but they can be found in virtually any part, such as in the tubers of potatoes and sweet potatoes. All of the protease inhibitors that have been isolated to date have proved to be proteins (Liener and Kakade, 1980). The nutritional significance of these inhibitors is that they are often associated with foods that are important sources of dietary protein, and they can potentially prevent the utilisation of this protein.

After ingestion, proteins are denatured by the acid in the stomach and subsequently hydrolysed by a number of enzymes in the small intestine to yield a mixture of amino acids and small peptides which can be absorbed. Inhibitors to most of these enzymes have been isolated, but it is trypsin that is by far the mostly widely affected, and this enzyme has also been the most closely studied.

The best known trypsin inhibitor occurs in soya beans. It has long been known that soya needs to be cooked in order for it to support growth in experimental animals. It was found that, in the case of soya, supplementing raw soya with methionine or cysteine (the limiting amino acids in soya) had the same effect, nutritionally, as cooking the soya. However, rather than being a specific effect on the absorption of these amino acids, it is believed that the inhibitor has a more general role of preventing or slowing protein digestion so that insufficient quantities of the limiting amino acids (whatever they happen to be) are absorbed. It also appears that other antinutrients may be present in raw soybean, and that the raw soya protein itself may be resistant to digestion (Liener and Kakade, 1980). As immature soya beans (green soya or edamame) are eaten as a vegetable in Japan and other countries, appropriate preparation techniques need to be considered in order to ensure that the product does not adversely interfere with normal digestion and absorption of nutrients.

From an industrial point of view, knowledge of the antinutritional characteristics of soya and other legumes is important in product design. Whilst many raw beans may not be intrinsically toxic, they may have reduced available protein levels compared with their cooked counterparts.

3.4.6 Cyanogenic glycosides

Many fruits and other plant foods contain compounds that have the potential to release cyanide. These compounds are usually glycosides - i.e. they consist of a sugar molecule linked to a cyanide group, usually indirectly through another component. The release of cyanide from these compounds occurs by enzymic hydrolysis, usually when the plant tissue is crushed or otherwise disrupted (allowing the active enzyme to reach the substrate), but it can also occur in the digestive system after the food has been eaten (Conn, 1969). Some plants are toxic because of their high levels of these compounds. Other foods we still consume, despite their containing moderate levels. The most well-known of these compounds is amygdalin, a cyanogenic glycoside first identified in bitter almonds, which on hydrolysis by an enzyme complex known as emulsin yields glucose, benzaldehyde and hydrogen cyanide. As a consequence, bitter almonds are generally considered to be inedible (Sadler *et al*, 1999), and commercial production is limited to the sweet variety in the USA (Conn, 1969). The other cyanogenic glycosides of potential importance in the human diet are dhurrin, which occurs in sorghum and other grasses, and linamarin and methyllinamarin, which are found in pulses (especially lima bean varieties) and cassava (Montgomery, 1980).

Cassava or manioc is a major basic food for large numbers of the world's population. It is the world's seventh largest food crop in terms of production area. The toxic potential of cassava has been known of for hundreds of years, and traditional methods of food preparation from cassava have been developed to reduce cyanide content. These include leaching out the linamarin precursor, washing in running water before cooking (bruising of the cassava root during harvesting often results in considerable cyanide release), and boiling in uncovered pots, so that the cyanide can evaporate. Fermentation steps also significantly reduce cyanogenic potential (Montgomery, 1980). The risk of food poisoning from cassava can be reduced by avoiding bruised or non-fresh material, and avoiding roots known to be bitter.

Box 30 - Cyanide formation in canned fruit

The cherry family, which includes plums, peaches, apricots and almonds, contain cyanogenic glycosides in their stones. In the canning of whole stone fruit (i.e. with kernels remaining), there is the risk that enough of the emulsin enzyme system will remain, resulting in amygdalin hydrolysis and cyanide release and accumulation. This is complicated by the fact that an excessive heat process will result in overprocessing and the fruit going mushy. The emulsin system in the stones of canned plums (termed beta-glucosidase, although it was subsequently found to consist of at least three enzymes working in sequence) was investigated by Haisman and Knight (1967). They found that the enzyme was thermally insulated to some extent. It was also stabilised by its isolation from the acidic constituents of the plum flesh, and by the relatively low water content and the high concentration of its substrate, amygdalin. After a canning process of 6 minutes at 100°C, cyanide content in the syrup rose to a maximum of 2ppm before declining; however, this level does not present a health hazard.

Hershkovitz and Kanner (1970) found that the enzyme system in the kernels of canned apricots remained active for long periods after processing, if the heat treatment was not sufficiently long to inactivate it. A process of 20 minutes at 86°C (which was inadequate to destroy the enzyme) resulted in hydrogen cyanide levels in the cans increasing to 16ppm after 150 days' storage, after which it remained practically constant. A process of 38 minutes at 86°C resulted in levels of only 1ppm, which did not change throughout storage. As well as the potential toxicity of the raised cyanide levels, canned apricots containing more than 6ppm hydrogen cyanide were unacceptable from a sensory point of view, having too strong an almond flavour.

Nowadays in the UK, most stone fruits are pitted before canning.

3.4.7 Toxins accumulated during growth

All plants and animals during their lifetime will accumulate various chemicals from their environment that are not specifically required by them and become assimilated more or less by accident. Some of these chemicals, if they are accumulated in high enough amounts, might be of toxicological significance to us when we eat the food.

Specific examples that are of concern at the moment are nitrates in leafy vegetables; heavy metals in various systems, and specific toxins in shellfish. In many cases, the best way to control levels of these unwanted substances is to control the environment in which the food is produced. However, this is generally a long-term control measure and more immediate steps have to be taken to protect human health. As most of the toxins can not always be 'processed out', the short term controls are usually based around the setting of maximum permitted levels, and the removal from the supply chain of food that does not meet the required standard.

Box 31 - Regulation of toxins in food

There have been limits on the levels of lead and arsenic in food for many years. The lead regulations (Anon, 1979) impose a 1mg/kg limit on food in general, including fruits and vegetables and most other 'primary' food products. There are higher limits for some products, e.g. fish (2mg/kg in general and 3mg/kg for canned fish), shellfish (10mg/kg), and tea (5mg/kg), and lower limits for other foods, e.g. infant foods (0.2mg/kg in general) and fruit juices (0.5mg/kg). Similar regulations for arsenic (Anon, 1959) also have a general limit of 1mg/kg for primary food products. These limits are set bearing in mind what is technically achievable, the volume of the product likely to be consumed and the target consumer (i.e. child or adult). The aim is to ensure that the health of the population is not affected by total lead consumption in food.

There has recently been concern over the levels of nitrates in some vegetables. Despite the application of good agricultural practice regimes in order to reduce the levels of nitrate in vegetables, the European Commission issued a regulation (EC, 1999) binding on all Member States fixing maximum limits for nitrate in spinach and lettuce. Depending on whether these were for sale as fresh or frozen produce, the method of growing (protected or open-grown) and the time of production (winter or summer), these values ranged from 2000 to 4500mg/kg.

Another area of concern is in toxin accumulation by fish and shellfish. In 1993, the European Commission issued a Decision, which fixed the maximum level of mercury in fishery products as 0.5 or 1.0mg/kg of fresh product, depending on the fish species. From time to time, levels of other toxins such as diarrhetic, paralytic and amnesic shellfish toxins give cause for concern and fishing bans in certain areas are imposed. These toxins are believed to be derived from the dinoflagellates that form the basis of the diet of the shellfish.

3.5 Summary

In summary, food is largely derived from other living organisms – mostly organisms bred for the specific purpose of providing food. The food harvested or derived from the plant or animal is entirely chemical in its composition. Most food chemicals are good for us or of no nutritional consequence, but some – a small minority – are potentially harmful and represent a hazard that has to be controlled. In some cases these hazards can be 'processed out', while in other cases their levels must be controlled in the raw material, raw material selection or their formation prevented during processing.

4. SPECIAL DIETARY NEEDS AND FOOD CHEMICAL COMPOSITION

There are many dietary problems that individuals may suffer from as a result of a genetic or immunological sensitivity or intolerance to individual chemicals in foods. This is different to most of the antinutrients or toxicants described in the previous chapter, which affect all people to a greater or lesser extent. Box 32 depicts the hierarchical classification of adverse reactions to food.

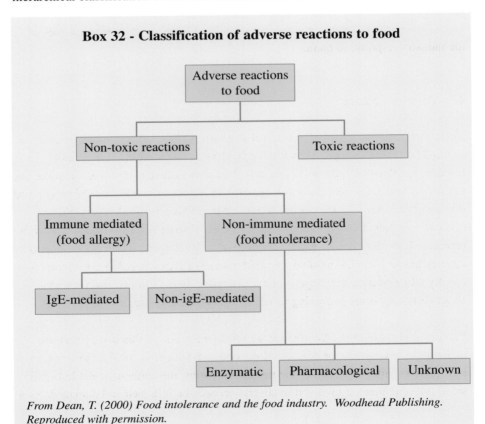

Box 32 - Classification of adverse reactions to food

From Dean, T. (2000) Food intolerance and the food industry. Woodhead Publishing. Reproduced with permission.

However, in both cases it is naturally occurring chemicals in the food which give rise to the 'problem'. Most people will be able to deal with the chemicals involved as part of their normal metabolism and with no indication of ill-effect. However, to those with genetic or immunological sensitivities, it is essential that they know from food labels which foods are suitable for them, and that the industry takes all reasonable precautions to ensure that what is indicated on the label is in fact a true reflection of the product.

The main areas where industry can help in such matters is in product reformulation to replace the component causing the reaction, clear labelling to indicate suitable foods where confusion might otherwise arise, and the prevention of cross-contamination at all points of the food supply chain using audit trails and quality assurance systems.

Some of the major dietary problems are explained below, with some examples of the industry response to them.

4.1 Coeliac disease

Coeliac diease, also called gluten-sensitive enteropathy or gluten intolerance, is induced by dietary wheat gliadins and related proteins found in other cereals, including spelt wheat - the wild ancestor of modern cultivars of wheat. Gliadins are one major fraction of the wheat storage protein, gluten (glutenins are the other), and it is the gluten which is the central factor in the production of wheat dough, and subsequently bread and similar products. Coeliac disease is a cell-mediated allergic reaction. These differ from typical hypersensitivity allergies, which are described later, in that they do not involve the rapid release of immunoglobulin E. Symptoms usually take significantly longer to develop, typically 6-24 hours, and peak after about 48 hours before beginning to subside (Taylor and Hefle, 2001).

The detailed aetiology of the disease is outside the scope of this book (there are many good descriptions of this, e.g. Troncone and Auricchio, 1991). However, in simple terms, after ingestion of the problem proteins, the epithelial cells lining the small intestine that are involved in absorption become inflamed and damaged. As a result, absorption of nutrients by the intestine is disrupted. This, along with the inflammation itself, results in a severe malabsorption syndrome characterised by diarrhoea, bloating, weight loss, anaemia, bone pain, chronic fatigue and weakness, and muscle cramps. In children, growth retardation and failure to gain weight are also seen. Coeliac disease is an inherited disease, but the inheritance is complex and poorly understood. Its incidence varies around the world, being highest in certain parts of Europe (as high as 1 in 300 in Galway, Ireland) (Kasarda, 1978). In the USA it affects about 1 in 3000, but is rare in those of Chinese or African descent (Taylor and Hefle, 2001).

The other major cereals that contain proteins that give rise to this effect are barley, triticale and rye. Maize and rice proteins are not a problem. The 'safety' or otherwise of oats has been investigated over many years, but recent research has suggested that oats can be eaten without harmful effect (Picarelli *et al*, 2001). However, the picture is not entirely clear (Thompson, 1997). Other work has indicated that oats do have a deleterious effect, and those suffering from the condition would need to take expert medical nutritional advice.

Many breakfast cereals are based on maize (corn) or rice, which are not gluten-containing cereals. However, the extruded breakfast cereal also contains a significant amount of malt, derived from barley, which itself contains a small amount of gluten-type protein. Although these breakfast cereals do not appear on the Coeliac Society's list of problem foods, it has been suggested that higher than normal consumption (4-5 bowls per day) might cause symptoms in some people, and again expert medical nutritional advice should be taken.

As stated above, the causative 'agent' in coeliac disease is a type of protein, and one which is fundamental to the formulation of a wide range of cereal-based, especially wheat-based, products. It is the physicochemical properties of gluten which give bread dough its characteristics, and these are impossible to mimic exactly using other cereal proteins. Thus, the industry has a three-pronged problem:

◆ to try and formulate some bakery products that are of acceptable quality and suitable for coeliac sufferers,
◆ to ensure that both suitable and unsuitable products are appropriately labelled, and
◆ to ensure that 'suitable' products are not contaminated with unsuitable ingredients.

Product labelling is a relatively straightforward issue (see box), although special care has to be taken to inform consumers in catering outlets. Suitable stock control and audit trails should avoid gross cross-contamination (e.g. using wheat flour instead of corn flour in a product). Analysis can also be used to confirm that the systems used are effective. Good hygiene and manufacturing practices will also reduce the risk of minor cross-contamination, which, although undesirable, does not constitute a health risk. Indeed the Codex Alimentarius Commission Code of Practice for 'gluten-free' labelling (Codex, 1994) allows for a small tolerance of gluten in the final product.

Product reformulation is a major technological problem. Wheat is a major ingredient in a variety of products, from breakfast cereals, through biscuits and cakes, to bread and pastries, and pasta. There are literally thousands of products which are not suitable for coeliac sufferers, and which would be candidates for reformulation with gluten-free flour and related ingredients. Not only does a product reformulation have to produce something that is edible and acceptable, it also has to have a reasonable shelf-life, and be economical to produce. The production runs are likely to be smaller, and unit costs will probably be higher, even after the development work has been completed. Despite this, the number of gluten-free alternatives that have been investigated are numerous, although many of these never reach the marketplace. Some products recently available through major retail outlets include:

- cake mixes which have rice flour and potato starch as principal ingredients;
- rice cakes;
- an organic bread based on maize flour and chestnut flour;
- a lemon cake made with gluten-free wheat starch;
- a selection of biscuits based on maize starch and flour, soya flour and tapioca starch

(from NewFoods, 2001; NewFoods, 2000; NewFoods, 1999).

Dean (2000) lists some wheat- and gluten-free products available on prescription and over the counter; these include pasta products, bread products and pizza bases, flour mixes and biscuits and crackers. The University of Saskatchewan (1999) recently patented an extruded pasta product based on pea flour, while Caperuto *et al* (2001) looked at the use of quinoa and maize flour mixtures; rice varieties have been examined for their suitability for making bread (Torres *et al*, 1999); and cassava flour has been evaluated for its use for muffin production (Chauhan *et al.*, 2001).

Box 33 - Food intolerance labelling

All packaged products must be labelled with a list of ingredients. Therefore those wishing to avoid one or more types of ingredients should be able to do so by reading the label. However, positive product labelling (e.g. 'suitable for coeliac sufferers' or 'gluten-free') is a significant marketing tool that can be both informative to the consumer and give a competitive edge to a product over a similar product over which there might be a doubt. Although most food intolerances occur in only a small percentage of the population, their immediate family and friends are also likely, in some instances, to buy and eat the 'suitable' product, rather than buy two separate products. However, excessive labelling of products (e.g. indicating that a product is 'gluten-free' if all products of that type are gluten-free) is not beneficial. The Codex Alimentarius Commission (Codex, 1994) say that a 'gluten-free' food shall be based on or contain:

◆ Gluten-containing cereals such as wheat, triticale, rye, barley or oats or their constituents, which have been rendered gluten-free; and/or

◆ Ingredients which do not contain gluten in substitution for the ingredients containing gluten which are normally used in food of that kind

---

4.2 Lactose intolerance

Lactose is the sugar found in milk, and is a disaccharide of glucose and galactose. Lactose intolerance arises due to insufficient levels in the small intestine of the enzyme beta-galactosidase (lactase), which is required to hydrolyse lactose to its constituent monosaccharides, which are then absorbed. This may be due to a long-term or progressive deficiency in the enzyme, or result from its loss due to another intestinal illness, such as gastroenteritis. In the latter case, once the original illness has passed, the lactose intolerance soon subsides. Lactase deficiency in adulthood is the norm for 70% of the world's population.

When lactose is not hydrolysed, it is not absorbed and passes into the colon, where the bacterial population metabolises it to carbon dioxide, water and hydrogen. Abdominal cramping, flatulence and frothy diarrhoea are the principal symptoms. The chronic condition tends to develop with increasing age, and varies greatly from one individual to another. The severity of the intolerance depends on how much or how little beta-galactosidase the individual has in his/her intestinal mucosa, and therefore how much lactose passes into the colon without being hydrolysed. Thus, some people can tolerate a moderate level of milk or milk products in the diet, whereas others have very low tolerance (Taylor and Hefle, 2001). The condition is unrelated to cow's milk allergies, which centre around an immune response to milk proteins (caseins). However, some of the product development to replace cow's milk in foods addresses both conditions.

The development of live yoghurts and similar products over the last 20 years has provided one route in for those with only moderate intolerance, as these products contain bacteria which produce beta-galactosidase. Lactose-hydrolysed milk is also available in the marketplace (Taylor and Hefle, 2001).

4.3 Phenylketonurea

This is a condition that results from the inability of some people to metabolise the amino acid phenylalanine correctly, and is the most common inborn error of metabolism (Whitney *et al*, 1998). Normally, phenylalanine is converted to tyrosine by the liver and to phenylpyruvic acid (a ketone body) by the kidneys. In some people, the liver enzyme is lacking, and excess ketone body formation occurs in the kidneys. This is toxic to the central nervous system and leads to brain damage. The condition can be controlled successfully by diets that restrict, but do not eliminate, phenylalanine intake - some phenylalanine is still required in the diet for protein synthesis. Protein-rich foods are excluded, as are bread and similar products which are made from flour with a high phenylalanine content. Specific formulations have been developed to provide the nutrients that sufferers would normally acquire in protein-rich foods.

One area in which food legislators and the food industry have had to act is in the labelling of food containing the sweetener aspartame, which has gained widespread

Box 34 - Development of soya milk

One of the main sources of milk substitutes that has been developed in the last 30 years in the West has been soya (Latin name *Glycine max*). Considering that soya is a legume related to kidney beans and similar vegetables, it is a testament to the ingenuity of both the modern food industry and their traditional Chinese counterparts that it is now regularly used as a milk substitute. Soya beans have been grown in China for more than 5000 years (Liu, 1997). During this time they have been used as a basis for fermented products, such as soy sauce and tempeh, as well as for non-fermented products like tofu. The high protein content of soya beans led to the development of soya-based texturised proteins that could be used as meat substitutes in product formulations suitable for vegetarians. These are now used in a very wide array of products for all sectors of the population. Soy milk itself is not new, and is believed to have first been made in China during the second century B.C. Basically, it is a water extract of ground soybeans, closely resembling dairy milk in both appearance and composition. The fact that its chemical composition is close to that of milk in several respects means that technologically it can be used like milk to a degree. Soya cream, soya yoghurt, and soya cheese (tofu) are all available as final products. Its advantage to those with milk protein allergy or lactose intolerance is that it contains neither. However, being so different to milk in its origins means that there are many subtle differences that require significant product development to overcome.

With soy milk itself, the main problem was the development of a beany, painty or rancid flavour, which was undesirable for the Western market. It was discovered that this characteristic flavour was due to the peroxidation of polyunsaturated fatty acids or esters, by lipoxygenase enzymes, to yield a variety of ketones, aldehydes and alcohols, which impart that flavour. For this to happen, the enzyme and its substrate need to be brought into contact in the presence of water. This only happens when the tissue is damaged (e.g. when the beans enter the grinding stage) - soaking of whole, sound, raw beans for 8-12 hours did not result in a beany flavour, and limiting the amount of water also prevented its development. However, the most efficient method for prevention of off-flavour development was enzyme inactivation: grinding the beans to a slurry at 80°C or above was effective. There are now several modern methods for making soymilk, involving either some form of enzyme denaturation or removal of the lipid substrate. Liu (1997) provides details on these and many other aspects of soya utilisation and processing.

use, especially in soft drinks. This is a dipeptide consisting of aspartic acid and phenylalanine, both of which occur in abundance in proteins. Although the amount of sweetener normally consumed only makes a small contribution to total phenylalanine intake, it is important that all sources that might be deemed to be 'unnecessary' are eliminated and therefore all products containing aspartame are labelled 'contains a source of phenylalanine'. In 1999 and 2000, CCFRA identified 224 new products in the UK marketplace which had this warning. Most were soft drinks, but there were examples of breakfast cereals formulated for the functional food market.

Box 35 - Favism

Favism is a haemolytic syndrome seen in susceptible individuals following the consumption of broad beans (fava beans - *Vicia faba*). Broad beans are cultivated and used as a human food throughout the world, but the disease is limited almost entirely to people originating from Mediterranean countries and the Middle East, with men being more susceptible than women. It results from a deficiency in the enzyme glucose-6-phosphate dehydrogenase in red blood cells (erythrocytes). This has a genetic basis, hence its specific geographical, racial and gender distribution. The end result, chemically, is a deficiency of reduced glutathione in erythrocytes, which is required for the stability of erythrocyte membranes. Lack of glutathione leads to internal haemolysis. The exact mechanism by which favism occurs, and the individual chemicals involved, is not fully understood, but it seems that derivatives of the nucleosides vicine and convicine may play a part, as may L-DOPA (3,4-dihydroxyphenylalanine) (Mager *et al*, 1980).

The disease is seen after the consumption of both cooked and raw beans, and the causative agents do not appear to be able to be 'processed out'. Avoidance would seem to be the only alternative for people with this enzyme deficiency. Sadler *et al* (1999) quote that Pythagoras may have suffered from favism. He forbade his followers from eating beans and was captured and killed by his enemies at the edge of a field of flowering broad beans, rather than trying to escape by running through the field. Inhalation of the pollen of broad bean plants is known to have a similar effect in susceptible individuals to eating the beans.

4.4 Diabetes

This is probably the most well-known metabolic disorder, and usually results from the deficiency or ineffectiveness of insulin, one of the two hormones that control the level of glucose in the blood. There are two types of diabetes. Type 1 is mainly due to an autoimmune-mediated destruction of the cells in the pancreas which produce insulin, and sufferers must have an external supply of insulin in order to control blood sugar levels. Type 2 diabetes, which accounts for 90% of diabetes worldwide, is caused either by abnormal insulin secretion or resistance to its effects. Many sufferers from this form can survive without external insulin, by careful control of their diet and/or by taking oral hypoglycaemic drugs (Zimmet *et al*, 2001, and subsequent articles in *Nature* 13th December 2001). It is type 2 diabetes that is mainly responsible for the significant increase in the number of sufferers worldwide, which is predicted to increase by 46% from 2000 to 2010.

Under normal circumstances, when blood glucose levels rise, as happens after a meal, insulin stimulates its removal from the blood. Individual cells in the body take up the amount they need, but both muscles and the liver can take up large amounts and convert it and store it as the polysaccharide glycogen. The liver can also convert glucose to fat. When blood glucose levels fall, as happens between meals, the hormone glucagon stimulates the conversion of liver glycogen back to glucose. With these two hormones acting together, a fairly narrow range of blood glucose concentration is maintained. In diabetes, either extra insulin has to be provided in the form of injections, or the intake of carbohydrate and its consequent effect on blood glucose levels has to be very closely monitored. Allowing blood glucose levels to 'stray' outside the normal range for long periods of time can be fatal. The glycaemic effect is the effect that consumption of a particular food has on blood glucose concentrations. This will vary depending on the mixture and volume of food consumed. Slow absorption, a modest rise in blood glucose and a smooth return to normal are desirable. The general rule of thumb is that foods high in free sugar will result in a more rapid and greater increase in blood glucose levels than those containing complex carbohydrates (e.g. starch), but this is not always the case, as other factors in the food (such as fat levels) will influence absorption rates.

It is now increasingly believed that most diabetics can control blood glucose levels by careful manipulation of a normal diet, without the need for special foods, and the market for these is diminishing. However, food items developed to be 'suitable for diabetics' have been available for many years. These are often formulated to be sugar- or glucose-free, but low- or reduced-carbohydrate versions of foods normally high in carbohydrate have also been formulated, to allow the diabetic a wider variety of choice and the chance to eat larger helpings of such foods (Bender, 1973).

Removing sucrose or glucose from a product or replacing them with alternatives is not straightforward, as they often have more than one role in the product. As described in a previous chapter, sugar has a significant preservative effect in some products, e.g. jams. Here, sorbitol, a sugar alcohol, can be used as a replacement. It has similar preservative properties, is slowly absorbed into the bloodstream, and is about 60% as sweet as sucrose. However, larger quantities (over 50g) have a laxative effect. Sugar also has a gelling effect in jams and jellies; sugar-free gels can be made from pectin with glycerol, but a separate sweetening source will be required, as glycerol has no sweetness. Fruit canned or bottled without sugar lacks viscosity, which has a detrimental effect on taste; thickeners such as pectin, sodium carboxymethylcellulose or seaweed extract (e.g. carageenan) can be added to correct this.

There has been much effort to reformulate chocolate and confectionery products to make them more accessible to diabetics. Fructose is slowly absorbed from the intestine and is rapidly metabolised, in the liver, without the involvement of insulin (Beckett, 1999) and is a valuable sucrose substitute in diabetic diets. As in other products, sorbitol has also been used in 'diabetic' chocolate.

4.5 Allergies

Food allergy has become one of the major concerns of the industry over the last 10 years or so. A detailed description of the allergic mechanism is outside the scope of this book, but in essence a true food allergy is an adverse reaction of the immune system to a substance (in the case of food, almost always a protein). There are two types of reaction: immediate hypersensitivity and delayed hypersensitivity.

One example of the latter is coeliac disease, which is described in Section 4.1. This section will look at immediate hypersensitivity reactions. These involve the production of allergen-specific immunoglobulin E antibodies when the individual is first exposed to the allergen. These result in the formation of sensitized cells which, on subsequent exposure to the allergen, release a wide range of allergic response mediators, of which histamine is one of the main players (Taylor and Hefle, 2001). Symptoms range from mild and annoying, to severe and life-threatening and include rashes, swellings, a precipitous drop in blood pressure and asthma and constriction of the airways. The severity will vary from one allergen to another and from person to person. It will also depend on the amount of the allergen eaten, previous levels of exposure, and other factors. However, one of the main features of allergic responses is the very small amount of allergen that is needed to trigger an effect, possibly even smaller in some cases than can be chemically detected.

There is no definitive list of which foods are or are not allergenic - it is possible that virtually every protein-containing food that we consume has the potential to be allergenic. However, experience has shown us that a limited number of foods affect a proportion of the population to a significant extent. Some people will react to one specific product only, while others may show a reaction to a number of (usually related) products. Although the number of people who are actually at serious risk from any particular allergen is small, the effects can be so extreme that the individual's immediate family and friends are also likely to avoid buying or using foods containing the ingredient in question. Thus, not only may food companies feel a moral obligation to enable sufferers to avoid problem ingredients, there are also strong commercial reasons.

Box 36 - Allergen lists

Many major food companies have drawn up their own lists of what they consider to be serious potential allergen-containing foods. Several learned and professional organisations have done likewise. Not surprisingly, these lists are not in complete agreement. There is no 'cut-off' line between allergic and non-allergic ingredients or foods, and the incidence of specific allergies varies remarkably from one culture to another, and even between countries within similar cultures (e.g. there is a relatively high incidence of allergies to celery in Scandinavia and France, whereas celery is not considered to be a significant problem in the UK). Recently, the Institute of Grocery Distribution (IGD, 2000) drew up a set of guidelines in which they 'amalgamated' the lists of serious potential allergens for food labelling purposes as given by several leading professional bodies. It consists of:

- peanuts;
- tree nuts;
- egg;
- fish;
- cows' milk;
- crustacea, molluscs and shellfish;
- sesame seeds; and
- soya beans.

Also given was a list products included under the tree nuts section. These included:

- almonds;
- Brazil nuts;
- cashews;
- chestnuts;
- hazelnuts;
- macademia nuts;
- pecans;
- pine nuts;
- pistachios; and
- walnuts.

There are great problems with defining a nut list - of the above, pine nuts are not botanically nuts, but seeds, and almonds are in fact the edible kernel of a soft fruit very closely related to apricots. However, the term nut is colloquially used to denote any hard fruit kernel, and the question frequently arises as to whether other nut-like products are significantly allergenic. As stated above, there is no clear cut-off point; the risk of an ingredient being allergenic is related to the chemical and physical structure of its proteins and the degree of exposure in the general population.

There are three ways in which the food industry can and does make very serious efforts to enable people with allergies avoid potentially dangerous foods. They are:

◆ clear labelling of products that contain the ingredients in question;

◆ reformulation of products, replacing allergenic ingredients with those considered to be not a problem; and

◆ prevention of cross-contamination of a 'non-allergenic' food with an allergenic ingredient.

Some product reformulation has taken place. Dairy-free products have been described earlier with reference to lactose intolerance. It is ironic that milk substitutes are usually based on soya, which is itself one of the major allergens. Coconut has historically been used in some formulations because it has a history of low allergenicity. More recently, efforts have been made to either remove nuts from products altogether or replace them with non-allergenic ingredients with the same functional properties (nuts are often used to give the desired texture and flavour to products such as curries and sauces).

Currently, food labelling regulations generally require that all ingredients of a food are listed, except if they form part of a compound ingredient (e.g. individual spices in a spice mixture of known composition). However, there is legislation being formulated at an EU level that will, in effect, require all ingredients that are significant potential allergens to be labelled. The list of such allergens is likely to be close to that drawn up by the Institute of Grocery Distribution (see box). Many food manufacturers are already carrying out labelling in the spirit of this legislation. There has also been much effort to inform consumers of when there is a risk of products not containing a specific allergen to have become contaminated with that allergen - the 'may contain traces of…' labelling. It must be said that some initial efforts in this respect have led to misunderstandings by consumers. The immediate response of a sufferer to a 'may contain…' label is often 'why may it contain… if it is not a deliberate ingredient?' The Institute of Grocery Distribution (IGD, 2000) has sought to redress this, and some major retailers are standardising on more informative explanations of the nature and degree of the risk.

In order to become acceptable, the above type of labelling, with proper explanations of why the risk exists, requires that all reasonable and practical precautions are taken to prevent cross-contamination. The unique problem with serious allergens is that minute levels of cross contamination may be a problem. Behind the simple statement 'all reasonable and practical precautions' lies a vast array of supplier checking, stock control, and hygiene measures that the industry has to take. In certain cases, an entire factory may be dedicated as free of a particular group of allergens, e.g. nuts.

Box 37 - Quality control procedures with respect to prevention of allergen cross-contamination

Although there are many different food products that can and do result in serious allergic responses in susceptible individuals, it is peanut and tree nut products that have received most attention in the UK. One particular problem from a food manufacturer's point of view is the many ways in which nuts are used. They can be used as whole nuts, small pieces or pastes, and oils, all of which can pose cleaning problems. Attention has also focussed increasingly on seeds such as sesame, sunflower and poppy. There seems little doubt that this type of ingredient has high allergenic potential, but as stated before, the scale of the potential problem will relate to the degree of usage of the ingredient. The comments that follow, taking nut ingredients as an example, could equally be applied to any other potentially allergenic ingredient that a manufacturer uses. These comments are by no means exhaustive, but are just to give examples of the type of factors that need to be considered.

The food processor has to decide how to treat any given group of potential allergens with respect to each other. For example, can all tree nuts be treated as one ingredient for the purposes of segregation and labelling, and can peanuts and seeds also be included in this? There is no reason why someone who is allergic to peanuts should be allergic to tree nuts or seeds, although some will; similarly, someone who is allergic to hazelnuts will not necessarily be allergic to Brazil nuts. Thus, if a food processor is dealing with more than one type of nut or seed product, it may be prudent to segregate them as much from each other as from other non-nut ingredients and products.

continued...

The first issue for food processors to address is to ensure that their ingredient suppliers are operating quality assurance and control procedures that are at least as good as their own. Segregation of nut from non-nut ingredients is significantly devalued if the incoming ingredients are already contaminated. Control of incoming stock is vital, through segregated warehousing and movement of ingredients. Correct labelling of ingredients is vital, along with adequate training of personnel so that formulation mistakes are not made.

Separate processing lines are the ideal manufacturing scenario; taken to the ultimate, this might involve a completely separate building, or a solid wall between the nut side and the non-nut side, with no connection between the two. However, in many situations this is commercially or physically impossible. It may be that there are too many combinations of allergens to make separate processing areas for one ingredient of value. Given this, the value of good design of processing lines can not be over-emphasised. Not only should they be designed to allow effective cleaning, but they should also inherently make it very difficult for direct cross-contamination from one line to another to take place. Work procedures should also be designed with this in mind. 'Rework' is a term used for recycling unused ingredients or left-over product back into the processing line. This is accepted and perfectly safe and hygienic practice, and is most associated with chocolate manufacture. Procedures must ensure that allergen cross-contamination can not occur through the incluson of nut-containing re-work in a nut-free line. Personnel working on nut lines should not swap to non-nut lines without taking the appropriate precautions. This would probably involve a complete change of outer clothes. Procedures would also have to be put in place to control the personal consumption of nut products on site by the workforce.

Cleaning and sanitation of processing lines and the associated environment is important in all food processing operations, but it assumes even greater importance if a nut-containing product is being followed by a nut-free line. Nut pastes are often used and it has been shown that these will contaminate subsequent products for a considerable period of time if not removed completely during a clean-down, i.e. they are not quickly flushed through (Hefle, 2000). One simple way to minimise this problem is to do the nut line last, before a scheduled shutdown and major cleaning operation. This may involve an extended run of the nut line. Introducing the allergenic component into the product as late as possible is also a useful strategy if it is feasible. Contamination issues will invariably arise from direct contact. Hefle (2000) suggests that there has never been a proven incident of cross-contamination from airborne particles.

continued...

Thought must be given to the type of information that will be available on the product label. If a company is only dealing in, say, walnuts, it will be much more informative to the allergic consumer to be told 'processed in a factory which processes walnuts' than 'may contain nuts'. Walnut allergy sufferers and their friends and family can then make a decision on whether to buy the product or not, and peanut and Brazil nut allergy sufferers can do so without any worries. However, if a company processes several different types of nuts, it may be difficult to come up with anything more than a general statement. A 'warning' type of label may still be valuable even if another related type of allergen is present in the product, e.g. sesame seeds.

Given the type of problems that allergen issues can cause, many manufacturers are looking at ways to minimise the use of some allergens. The most obvious is to use non-allergenic alternatives if they will do the same job. Protein-free artificial or nature-identical flavours can be used without allergy concerns. Also, if the allergenic ingredient is being used in very small amounts, it may be possible to omit it from the formulation altogether. In contrast, if an allergenic ingredient is already being used in several other process lines, it may be best to continue using this one rather than an alternative allergen. A related strategy is to avoid new ingredients with high allergenic potential, e.g. cottonseed, poppy seeds and sunflower seeds. Although there is little evidence of these being a problem at the moment, their increased use and a greater exposure of the population to them may result in allergies being reported with greater frequency.

One specific issue that frequently arises is the use of nut oils. Highly refined oils should not contain any protein, and therefore will not pose a problem. However, guaranteeing the refined nature of the oil, given the significant volumes of oil that might be used in a formulation or process and the small amount of protein that might cause a problem, is probably one of the most difficult tasks of a processor. Having 'highly refined peanut oil' on the list of ingredients on a product label is probably going to guarantee that a peanut allergy sufferer will not buy it.

Finally, if a very small percentage of a company's product turnover involves nuts, they may decide to delete these lines completely. This may happen, for example, if a company has a special seasonal run, e.g. of Christmas cake containing nuts, but does not handle nuts for the rest of the year. In this situation, a nut-free policy is probably easier to manage than all the precautions that otherwise would have to be taken.

5. FOOD LABELLING AND CHEMICAL COMPOSITION

The labelling of food products within the EU is highly regulated. The regulations and associated guidelines and codes of practice set out to ensure that the consumer has the opportunity to know as much as possible about what is in the product being bought. As far as food chemical composition is concerned, there is legislation or guidance covering the display of nutrition information, the making of claims (e.g. high in fibre, low in fat), the description of flavourings, the use of the word 'natural', and indications of vitamin and mineral content. There are also some specific requirements to indicate if a food contains a chemical that may be a health risk to a section of the population, and further requirements have been proposed.

It is not possible to detail the entire regulations relating to this area, although the following is an attempt to explain the main areas and to give an indication of the range of legislation or guidance that exists in the UK. It has also been attempted to describe it in simple terms. As a result, certain exceptions will not have been covered, and it is important that the original legislation and/or an authoritative source is contacted over specific points of law.

5.1 Nutrition labelling

It is not a legal requirement to label products with the nutritional content of the food, except if a nutritional claim is being made about the food. However, it is probably true to say that a combination of the desire of industry to provide such information, and consumer pressure to receive it means that nowadays most food labels do contain such information. Nutrition labelling is laid out in the Food Labelling Regulations Statutory Instrument 1996 Number 1499. The regulations and subsequent amendments are available on the HMSO website, and compilations of these and other food legislation are published by several organisations (e.g. CCFRA, 2001).

When displaying nutrition information, the minimum declaration is of the amounts of energy, protein, carbohydrate and fat per 100g or per 100ml of product as consumed. This must be displayed conspicuously on the label and in tabular form. If more information is provided, the amounts of sugars, saturates, salt and fibre must be given, along with any subcomponents of these for which a claim is made. Thus the two main formats are:

Table 7 - Nutrition labelling formats

Group 1 (also known as the 'big 4')

Energy	(in kiloJoules (kJ) and kiloCalories (kcal))
Protein	(in grams)
Carbohydrate	(in grams)
Fat	(in grams)

Group 2 ('big 4 + little 4')

Energy	
Protein	
Carbohydrate	
of which:	
sugars	(in grams)
Fat	
of which:	
saturates	(in grams)
Fibre	(in grams)
Sodium	(in grams)

It is not permissible to mix and match the above, and if a nutrition claim is being made about sugars, saturates, fibre or sodium, then the Group 2 format must be used. It is also permitted to declare the levels of polyols, starch, mono- and polyunsaturates (all in grams), cholesterol (in milligrams) and vitamins and minerals (in milligrams or micrograms as prescribed) present. (For declaring vitamin and mineral levels, significant quantities must be present - i.e. 15% of RDA in 100g). These can be included in either the Group 1 or Group 2 format, and if a claim is made about one or more of these then it/they must be included in the nutrition declaration. The level of saturates must be declared if mono- or polyunsaturates or cholesterol are declared.

The Institute of Grocery Distribution has also suggested that key data on a food's energy and fat contents should be highlighted.

The above are the basic requirements of nutrition labelling, but additional declarations can be given per quantified serving or portion.

Amounts given in nutrition information are usually the amounts contained in the food as sold. However, where sufficiently detailed instructions are given for the preparation of a food for consumption, they may be the amounts contained in the food after completion of the preparation instructions.

Box 38 - 'Per serving' nutrition labelling

One interesting example of per-serving nutrition labelling of food is often seen with breakfast cereals, where per serving details are given inclusive of a typical amount of milk that would be consumed with the cereal. This is important as, although cereals themselves are low in fat, the total fat intake during the meal is increased if full-fat or semi-skimmed milk is added. Legally, this has implications as to whether and how a low-fat claim could be made with such a product. If the product, as typically consumed, did not meet the guidance for 'low-fat', even though the dry cereal did, it might be seen as misleading if this was not clearly pointed out on the label.

5.2 Nutrition claims

There are specific restrictions about what can be claimed about the nutritional properties of a food or its ingredients. It is not permitted to claim that a food can prevent, treat or cure a human disease. Claims can be made that a food is high or low in a specific component. Some of these claims are regulated by law, others are controlled by guidelines that were issued by the Ministry of Agriculture, Fisheries and Food (MAFF), following recommendations from the Food Advisory Committee (FAC). Examples of the form which these claims must take are: 'source of vitamin

X', 'excellent source of protein', 'low in fat' or 'reduced energy content'. If a food type is typically low or high in a particular component, then it must not be implied that it is 'better' than other similar foods, but it can be labelled, for example, as 'a low-fat food' or 'a high-fibre food' as appropriate. The requirements for such claims are detailed below.

Proteins:

The Food Labelling Regulations state that to justify a claim that a food is a source of protein, a normal daily portion of that food must contribute at least 12g of protein. Also, at least 12% of the energy content of the food must be supplied by the protein. If it is described as an excellent or rich source of protein, then the protein must contribute at least 20% of the energy content.

Energy value:

The Food Labelling regulations state that a food can be described as having a reduced energy value, if that energy value is no more than three quarters of the energy value of a similar 'standard' food. To be described as having a low-energy value, it must provide no more than 167kJ (40kcal) per 100g. A low-calorie drink must contain no more than 42kJ (10kcal) per 100ml.

Fats:

Current guidelines based on UK FAC recommendations require that, for a food to be described as being a low-fat food, both a normal sized serving and 100g of the product must contain no more than 3g of fat. A food described as 'reduced-fat' should have a fat content that is at least 25% less than a standard version of the product. A 'fat-free' food should contain no more than 0.15g fat per 100g.

Claims are often made for saturates in a food. FAC guidelines suggest a maximum figure of 1.5g per 100g and per normal serving for a 'low-saturates' claim and no more than 0.1g per 100g for a 'saturates-free' claim. As with total fats, a 'reduced-saturates' claim should indicate at least a 25% reduction in the level of saturates over the standard product.

Box 39 - Butter and margarine - can you tell the difference?

Margarine was patented and first manufactured by a French chemist, Hippolyte Mege Mouries, in 1869. The driving force behind its original development was to meet the increasing demand for a butter-like spread, which could not be met in urban areas by butter itself. The original process was designed to mimic the production of butterfat by the cow and was based largely on tallow, skim milk, and cow udder tissue. Subsequently, some margarines were produced from lard or unfractionated beef suet with liquid vegetable oils added to reduce the overall melting point. In the early 1900s, some 100% vegetable oil margarines were developed. This involved hydrogenating the unsaturated fatty acids to yield a solid product with a high proportion of saturated fats, similar to that found in butter. Advertising campaigns in the 1960's in the UK concentrated on convincing the consumer that margarine could substitute for butter in terms of flavour and texture. In the USA, consumption of margarine exceeded that of butter by about 1960 (Hui, 1996).

In the UK, margarines that did not look like direct butter substitutes became more popular during the 1970s. This trend, which has accelerated over the past 30 years, has been fueled by both nutritional concerns and a desire for a different type of convenience product, specifically one which can be spread 'straight from the fridge'. The two demands have been able to be met by a series of products, packaged in plastic tubs, which are based on emulsions with a relatively high water content, and a significant reduction in the degree of hydrogenation. This has resulted in margarines with a low content of saturated fatty acids and trans-unsaturated fatty acids (both with a poor health reputation, currently), and higher levels of cis-unsaturated fatty acids (the major form naturally occuring in both plants and animals).

Sugars:

Current guidelines state that for a 'low-sugar' or 'low-sugars' claim, both a normal serving and 100g should contain no more than 5g of sugar(s). A sugar-free food should contain no more than 0.2g per 100g. In addition, a 'no added sugar' claim can be made if no sugar or foods mainly composed of sugar have been added to the food or to any of its ingredients.

Salt/Sodium:

The health concerns about salt are due to the sodium content of salt. Current guidelines indicate that a 'low-sodium/salt' claim can be made for a food if both a normal sized portion and 100g of the food both contain no more than 40mg of sodium. 'Salt/sodium-free' foods should contain no more than 5mg sodium per 100g. A 'no-added-salt' claim can be made if no sodium salts have been added to the food or any of its ingredients. Thus, a 'no-added-salt' claim cannot be made if, for example, monosodium glutamate (MSG) has been added to the product.

Box 40 - Salt and sodium

Although salt is the major source of sodium in the diet, it is by no means the only source. As well as the fact that all foods contain sodium in various amounts, it is also added as part of several permitted additives, notably monosodium glutamate. However, it is sodium chloride (table salt) that contributes the majority of sodium in the western diet.

Fibre:

Current guidelines state that for a 'source of fibre' claim, a food must contain at least 3g of fibre per 100g or in the reasonable expected daily intake of the food (if this is less than 100g). For a claim to be made that it is an 'increased' source of fibre, this condition must be met, and it must contain at least 25% more than a similar, 'standard' food. For a food to be claimed as a rich source of fibre, it must contain at least 6g per 100g or per serving.

Vitamins and minerals:

Claims relating to vitamins and minerals are laid out in the Food Labelling Regulations, and relate to the Recommended Daily Allowance (RDA) for each vitamin and mineral. Many different authorities have published opinions on RDAs and similar terms for mineral and vitamin requirements (e.g. estimated average requirement, and Reference Nutrient Intake - see DoH, 1991b). As described earlier, the recommended requirements often relate to total energy intake (i.e. someone with a greater caloric intake will require more of certain vitamins). Therefore, RDAs are quoted in the Food Labelling Regulations and these must be used for any claims made. In addition, claims can only be made for the vitamins and minerals listed.

To claim that a food is a rich or excellent source of a vitamin or mineral, the normal daily intake of that food must provide at least half of the RDA of the named vitamin or mineral. For any other claim for vitamins and minerals (e.g. 'fortified with'), a normal daily serving of the food must supply at least one sixth of the RDA.

Table 8 - RDAs for vitamins for which claims can be made

Vitamin	RDA
Vitamin A	800 ug
Vitamin D	5ug
Vitamin E	10mg
Vitamin C	60mg
Thiamin	1.4mg
Riboflavin	1.6mg
Niacin	18mg
Vitamin B6	2mg
Folic acid/folacin	200ug
Vitamin B12	1ug
Biotin	0.15mg
Pantothenic acid	6mg

Table 9 -RDAs for minerals for which claims can be made

Mineral	RDA
Calcium	800mg
Phosphorus	800mg
Iron	14mg
Magnesium	300mg
Zinc	15mg
Iodine	150ug

Box 41 - Folic acid claims

In 1997, the Health Education Authority (HEA) launched a campaign to raise the awareness of the importance of folic acid in the prevention of neural tube defects in the unborn baby. As part of this campaign, the HEA developed 2 flashes to highlight foods fortified with folic acid, namely: 'contains folic acid' and 'with extra folic acid'. The 'contains folic acid' flash constitutes a source claim under the UK Food Labelling Regulations and the product would be required to contain one sixth of the RDA, i.e. 33ug, in an amount a person could reasonably be expected to consume in one day. If the 'extra folic acid' flash were used, this would be equivalent to claiming the food as a rich or excellent source claim under the regulations and the product would need to contain 100ug in an amount a person could reasonably be expected to consume in one day.

The HEA Flash scheme relates to the synthetic form of folic acid (free pteroyl glutamic acid), which is only found in fortified foods.

Subsequent guidance from LACOTS (the Local Authority Co-ordinating Body on Food and Trading Standards now LACORS the Local Authorities Co-ordinators of Regulatory Services) suggested that other wording could be used on product labels, without infringing the prohibition of medicinal claims in the Food Labelling Regulations. These included the following:

♦ Important: doctors recommend that women trying to become pregnant, and in the first 12 weeks of pregnancy, take an extra 400ug supplement of folic acid a day for the normal development of the baby's spinal cord.

♦ Folic acid contributes to the normal growth of the foetus/unborn baby/baby in the womb.

♦ Folic acid is good for foetal development/development of the foetus.

5.3 Use of the word 'natural' and similar phrases

The meaning of the word 'natural' and phrases such as 'free from artificial x', as they have often been used in the context of food, can be ambiguous and in 1990 the UK FAC produced revised guidance to try and ensure that consumers were not misled by them (this was reviewed in 2001). Much of the guidance is outside the scope of this book. However, there are certain aspects of this guidance that are relevant, particularly as they apply to additives and flavourings. Both of these groups of chemicals can be extracted from biological sources, or synthesised chemically.

The guidance states that the word 'natural' should only be used with regard to additives if they have been obtained from recognised food sources by appropriate physical processing (including distillation and solvent extraction) or by traditional food preparation processes. A similar caveat applies to flavourings described as natural, and is embodied in legislation in EC Directives 91/71/EEC and 91/72/EEC.

The term 'nature identical' can be used to describe a flavouring or additive that has been chemically synthesised, but is identical to that occurring in nature. However, only defined chemicals can be described as nature identical: it is generally not

---

Box 42 - Flavour and flavoured

When using an artificial flavouring, can the word 'flavoured' be used as part of the name of the food?

If the flavour of the named ingredient is not derived wholly or mainly from the natural source - in other words it is derived from an artificial or nature-identical flavouring - the word **'flavour'** must be used. For example, 'strawberry flavour mousse'.

However, when the named ingredient or natural flavouring is used the word **'flavoured'** can be used - as in 'strawberry flavoured mousse'. Alternatively, where the natural ingredient or natural flavouring is used, it is also permissible to omit the word 'flavoured' altogether - as in 'strawberry mousse'.

possible to describe strawberry flavouring as nature identical, as strawberry flavour is derived from a cocktail of many dozens of compounds which occur in different amounts and it would be necessary to mimic this cocktail exactly.

5.4 Warnings and similar indications on labels

There are some indications that have to be put on labels if certain additives or ingredients are present. Those controlled by the UK Food Labelling Regulations are as follows:

Sweeteners: a food containing authorised artificial sweeteners must be labelled with the indication 'with sweetener(s)'. If added sugar(s) are also present, then the indication must take the form 'with sugar(s) and sweetener(s)'.

Aspartame: a food containing aspartame must be labelled with the indication 'contains a source of phenylalanine'.

Polyols: a food containing more than 10% added polyols must be labelled with the indication 'excessive consumption may produce laxative effects'.

Industry also labels products with regard to suitability for vegetarian, vegan, milk-free, and gluten-free diets, and also as regards diabetes and allergies, especially to nuts and nut derivatives.

6. CONCLUSIONS

Food is comprised entirely of chemicals. Some of these provide our energy and body mass needs or help with metabolic processes, others make the food palatable, while others are undesirable to one degree or another. Many have no role to play at all; these and the undesirable compounds are merely associated with the food in its raw or processed state. A relatively small number of chemicals are added to food during formulation or processing, but most of these have no dietary significance at all.

In addition to requiring convenience and safety from food products, today's consumer also demands good eating quality (e.g. good flavour) and specific dietary attributes, such as more fibre and protein, lower fat levels, and the complete absence of some chemicals (e.g. nut proteins, lactose or wheat gluten). This impinges on the whole of the food manufacturing chain, from the growing of fruit, vegetable and cereal crops and the production of livestock, through the development of completely new foods such as Quorn, to the formulation of an unprecendented range of processed and manufactured foods. The production of many of these foods has required many years of development, and some tasks are still proving to be a problem - such as the reduction of fat levels in certain bakery products. Not only have food technologists been involved in these developments, but also the demand for products suitable for consumers with specific medical conditions, such as severe allergies or intolerances to certain ingredients has meant that the whole management of the food supply chain has been under scrutiny.

All this has occurred against the backdrop of media scare stories on certain aspects of food production and composition, some of which have diverted attention away from many of the more important aspects of the dietary role of food in our lives. It has been the aim of this book to draw attention to the chemical nature of food and to put into some perspective the relative importance of the different groups of chemicals that we consume in our food.

7. RELEVANT ORGANISATIONS AND ASSOCIATIONS

Biochemical Society
Tel: 020 7580 5530
Fax: 020 7637 7626
E-mail: genadmin@biochemistry.org
website: www.biochemistry.org

British Dietetic Association
Tel: 0121 616 4900
Fax: 0121 616 4901
e-mail: bda@dial.pipex.com
website: www.bda.uk.com

British Nutrition Foundation
Tel: 020 7404 6504
Fax: 020 7404 6747
e-mail: postbox@nutrition.org.uk
website: www.nutrition.org.uk

Coeliac Society
Tel: 01494 437278
Fax: 01494 474349
Website: www.coeliac.org.uk

Department of Environment,
Food and Rural Affairs
Tel: 020 7238 6000
Fax: 020 7238 6591
website: www.defra.gov.uk

Department of Health
Tel: 0207 210 4850
e-mail: dhmail@doh.gsi.gov.uk
website: www.doh.gov.uk

Diabetes UK
(formerly British Diabetic Association)
Tel: 020 7323 1531
Fax: 020 7637 3644
website: www.diabetes.org.uk

Food Standards Agency
Tel: 020 7276 8000
website: www.food.gov.uk

Institute of Food Science and Technology
Tel: 020 7603 6316
Fax: 020 7603 6317
e-mail: info@ifst.org
website: www.ifst.org

Royal Society of Chemistry
Tel: 020 7437 8656
Fax: 020 7437 8883
e-mail: rsc1@rsc.org
website: www.chemistry.rsc.org/rsc/

8. GLOSSARY

The majority of terms relating to food chemistry are explained throughout the text of this book, but the following general definitions may be of use.

Diet - a description of the overall make-up of the food that a person consumes over a period of time.

Functional food - a food that beneficially affects one or more target functions in the body, beyond adequate nutrition, in a way that improves health or well-being or reduces the risk of disease.

Health - the state of well-being or otherwise of an individual. Only people have health; individual foods cannot be termed healthy, although the terms healthy and unhealthy can be used to describe an overall diet.

Nutraceutical - a chemical in or added to a food which is not a nutrient, but which has potential beneficial effects.

Nutrient - a term sometimes used to describe essential dietary factors such as vitamins, minerals, fatty acids and amino acids, but in this book taken to also include sources of energy (fats and carbohydrates) and other components of food such as fibre which potentially have a benefit on health.

Nutrition - the process by which living organisms take in and use food for the maintenance of life, including energy production and growth.

Recommended Daily Allowance - the amount of nutrient (typically a vitamin or mineral) intake that is greater than the requirements of about 97.5% of the population; i.e. greater than the average requirement of the population, allowing for individual variation. Also called Reference Nutrient Intake. In contrast the Estimated Average Requirement is the average requirement of the population; assuming a normal distribution of needs, the EAR will satisfy half of the population.

Toxin - a chemical in a food, either naturally present, a contaminant or deliberately added, that has the potential to have an adverse effect on health.

9. REFERENCES

Anon (1959). The Arsenic in Food Regulations 1959, as amended. HMSO

Anon (1979). The Lead in Food Regulations 1979, as amended. HMSO

Anon (1996). The Food Labelling Regulations 1996. SI 1499, as amended. HMSO

Anstis, J. and Cauvain, S.P. (1998) The effects of sugar and alternatives in cakes. CCFRA R&D Report No. 68

Ayhan, Z., Yeom, H.W., Zhang, Q.H. and Min, D.B. (2001). Flavor, color and vitamin C retention of pulsed electric field processed orange juice in different packaging materials. Journal of Agricultural and Food Chemistry, **49** (2): 669-674

Bacchem, C.W.B., Speckmann, G.J., van der Linde, P.C.G., Verheggen, F.T.M., Hunt, M.D., Steffens, J.C. and Zabeau, M. (1994) Antisense expression of polyphenol oxidase genes inhibits enzymatic browning in potato tubers. Bio/Technology 12 1101-1105.

Baker, R.T.M. (2002) Canthaxanthin in aquafeed applications: is there any risk? Trends in Food Science and Technology 12 (7): 240-243.

Banks, J.M., Hunter, E.A. and Muir, D.D. (1993). Sensory properties of low fat Cheddar cheese: effect of salt content and adjunct culture. Journal of the Society of Dairy Technology, **46** (4): 119-123

Beckett, S.T. (1999). Industrial Chocolate Manufacture and Use. Publ: Blackwell Science

Bender, A.E. (1973). Nutrition and Dietetic Foods. Publ: Leonard Hill Books

Bender, D.A. and Bender A.E. (1999). Benders' Dictionary of Nutrition and Food Technology. 7th edition. Publ: Woodhead Publishing

British Nutrition Foundation (1994). Food Fortification. BNF Briefing Paper

British Nutrition Foundation (2001). Selenium and Health. BNF Briefing Paper.

Caperuto L.C., Amaya-Farfan J., Camargo C.R.O. (2001) Performance of quinoa (*Chenopodium quinoa* Willd) flour in the manufacture of gluten-free spaghetti. Journal of the Science of Food and Agriculture, **8** (1): 95-101.

Catterall, P.F. (2001) Fat replacers and substitutes. Challenges in the developments of reduced fat bakery products. In: "New Technologies - Future Today" CCFRA Symposium Proceedings.

Cauvain S.P., Hodge D.G. and Screen A.E. (1988). Changes in the fat component of cakes and biscuits to meet dietary goals. Part 1. Fat reduction in cakes. FMBRA Research Report No.140.

CCFRA (1999) New Technologies Bulletin No. 19

CCFRA (2001). UK Food Law Notes.

Chauhan, S., Lindsay, D., Rey, M.E.C., and von Holy, A. (2001). Microbial ecology of muffins baked from cassava and other nonwheat flours. Microbios, **105**, (410): 15-27.

Codex Alimentarius Commission (1994). Codex Alimentarius Volume 4. Codex Standard for "gluten-free foods". Standard 118-1981

Conn, E.E. (1969). Cyanogenic glycosides. Journal of Agricultural and Food Chemistry, **17** (3): 519-526

Day, B. (2001) Fresh prepared produce: GMP for high oxygen MAP and non-sulphite dipping. CCFRA Guideline No. 31 - Campden & Chorleywood Food Research Association.

Dean, T. (2000). Food intolerance and the food industry. Publ: Woodhead Publishing.

Department of Health (1991a). The fortification of yellow fats with vitamins A and D. Report on Health and Social Subjects 40.

Department of Health (1991b). Dietary reference values for food energy and nutrients for the United Kingdom. Report on Health and Social Subjects 41.

Department of Health (2000). Folic acid and the prevention of disease. Report on Health and Social Subjects 50. Committee on the Medical Aspects of Food and Nutrition Policy

European Commission (1993). Commission Decision of 19 May 1993 determining analysis methods, sampling plans and maximum limits for mercury in fishery products.

European Commission (1999). Commission Regulation (EC) No 864/1999 of 26 April 1999 amending Regulation (EC) No 194/97 setting maximum levels for certain contaminants in foodstuffs.

European Parliament (1997). Regulation (EC) No. 258/97 of the European Parliament and of the Council of 27 January 1997 concerning novel foods and novel food ingredients. Official Journal of the European Communities L43 v40 14 February p. 1

Dyke, S.F. (1965). The chemistry of the vitamins. Publ: Interscience Publishers

Fassett, D.W. (1973). Oxalates. In: Toxicants Occurring Naturally in Foods. 2nd Edition. Publ: National Academy of Sciences

Fast, R.B. and Caldwell, E.F. (1990). Breakfast cereals and how they are made. Publ: American Association of Cereal Chemists

Garrow, J.S. and James, W.P.T. (1993). Human Nutrition and Dietetics. 9th Edition. Publ: Churchill Livingstone

Gibson, G.R. and Williams, C.W. (2000). Functional Foods: Concept to Product. Publ: Woodhead Publishing

Golovan, S *et al* (2001). Pigs expressing salivary phytase produce low-phosphorus manure. Nature Biotechnology, **19**: 741-745

Haisman, D.R. and Knight, D.J. (1967). Beta-glucosidase activity in canned plums. Journal of Food Technology, **2**: 241-248

Hefle, S. (2000). Food allergens and GMP. Lecture given at Leatherhead Food Research Association Conference 'Food Allergy - new Insights and Directions'.

Hershkovitz, E. and Kanner, J. (1970). The effect of heat treatment on beta-glucosidase activity in canned whole apricots. Journal of Food Technology, **5**: 197-201

Holland, B., Welch, A.A., Unwin, I.D., Buss, D.H., Paul, A.A. and Southgate, D.A.T. (1991). McCance and Widdowson's Composition of Foods. 5th edition. Royal Society of Chemistry and Ministry of Agriculture, Fisheries and Food

Honein, M.A., Paulozzi, L.J., Matthews, T.J., Erickson, J.D. and Lee-Yang, C.W. (2001). The effect of folic acid fortification of the US food supply on the occurrence of neural tube defects. Journal of the American Medical Association, **285**: 2981-2986

Hui, Y.H. (1996). Bailey's Industrial Oil & Fat Products. Volume 3. Edible Oil and Fat Products: Products and Application Technology. Publ: John Wiley

Institute of Grocery Distribution (2000). "This product may contain nuts". Voluntary labelling guidelines for food allergens and gluten.

Jaffe, W.G. (1973). Toxic proteins and peptides. In: Toxicants Occurring Naturally in Foods. 2nd Edition. Publ: National Academy of Sciences

Jaffe, W.G. (1980). Hemagglutinins (lectins). In: Toxic Constituents of Plant Foodstuffs. (ed. Liener, L.E.) 2nd Edition. Publ: Academic Press

Jones, J.L., Roddick, J. and Smith, D. (1996). Potatoes: the poison potential. Food Manufacture, **71**(11): 36-37

Kasarda, D.D. (1978). The relationship of wheat proteins to celiac disease. Cereal Foods World, **23** (5): 240-244

Langseth, L. (1995). Oxidants, antioxidants and disease prevention. International Life Sciences Institute Monograph

Lee, N.K., Yoon, J.Y. and Lee, S.R. (1995). Changes in heavy metals and vitamin C content during storage of canned and bottled orange juices. Korean Journal of Food Science and Technology, **27** (5) 742-747

Lee, H.S. and Coates, G.A. (1999). Vitamin C in frozen, fresh squeezed, unpasteurized, polyethylene-bottled orange juice: a storage study. Food Chemistry, **65** (2): 165-168

Liener, I.E. (1980). Toxic Constituents of Plant Foodstuffs. 2nd Edition. Publ: Academic Press

Liener, I.E. and Kakade, M.L. (1980). Protease inhibitors. In: Toxic Constituents of Plant Foodstuffs. 2nd Edition. Ed: I.E. Liener Publ: Academic Press

Lian, Y.S., Liang, G.H., Deng, X.Q., Dong, G., Chen, X.H. and Xie, L. (2000). Stability of vitamin C in beverage of orange juice. Food Industry, (2): 14-15

Liu, K. (1997). Soybeans: Chemistry, Technology and Utilization. Publ: Chapman and Hall

Llewellyn-Davies, D. (2001). Focus on new 'healthy eating' products. CCFRA Marketplace Report

Macedo, A.C. and Malcata, F.X. (1997). Technological optimisation of the manufacture of Serra cheese. Journal of Food Engineering, **31**(4): 433-447

McEwan, J.A. (1999). Barriers to the consumption of reduced fat bakery products: a consumer approach. CCFRA R&D Report 80

McEwan, J.A. and Clayton, D. (1999). Barriers to the consumption of reduced fat bakery products: a qualitative approach. CCFRA R&D Report 78

McEwan, J.A. and Sharp, T.M. (1999). Barriers to the consumption of reduced fat bakery products: final report. CCFRA R&D Report 85

Mager, J., Chevion, M. and Glaser, G. Favism. (1980). *In*: Toxic Constituents of Plant Foodstuffs. 2nd Edition. Ed: I.E. Liener Publ: Academic Press

Massaioli, D. and Haddad, P.R. (1981). Stability of the vitamin C content of commercial orange juice. Food Technology in Australia, **33** (3): 136,138

Ministry of Agriculture, Fisheries and Food (1999). 1997 Total diet study: aluminium, arsenic, cadmium, chromium, copper, lead, mercury, nickel, selenium, tin and zinc. Food Surveillance Information Sheet No. 191

Mistry, V.V. and Kasperson, K.M. (1998). Influence of salt on the quality of reduced fat Cheddar cheese. Journal of Dairy Science, **81** (5): 1214-1221

Montgomery, R.D. (1980). Cyanogens. In: Toxic Constituents of Plant Foodstuffs. 2nd Edition. Ed: I.E. Liener Publ: Academic Press

National Research Council (USA) (1973). Toxicants Occurring Naturally in Foods. 2nd Edition. Publ: National Academy of Sciences

NewFoods (2001). A CD-RoM database of new products purchased in the UK 1999-2000

NewFoods (2000). A CD-RoM database of new products purchased in the UK 1998

NewFoods (1999). A CD-RoM database of new products purchased in the UK 1997

Oberleas, D. (1973). Phytates. In: Toxicants Occurring Naturally in Foods. 2nd Edition. Publ: National Academy of Sciences

Orchard House Foods (2001). Trade press release of 4th July 2001 on extreme heat processed orange juice. www.ohf.co.uk 12/12/01

Paul, A.A., Southgate, D.A.T. and Russell, J. (1980). McCance and Widdowson's Composition of Foods. 1st Supplement to 4th edition. Ministry of Agriculture, Fisheries and Food

Picarelli, A., Di tola, M., Sabbatella, L., Gabrielli, F., Di Cello, T., Anania, M.C., Mastracchio, A., Silano, M., De Vincenzi, M. (2001). Immunologic evidence of no harmful effect of oats in celiac disease. American Journal of Clinical Nutrition, **74**: 137-140.

Ranken, M.D., Kill, R.C., and Baker, C.G.J. (1997). Food Industries Manual. 24th Edition. Blackie Academic and Professional.

Sadler, M.J. (1988). Quorn. Nutrition and Food Science, **112**: 9-11

Sadler, M.J., Strain, J.J. and Caballero, B. (1999). Encyclopaedia of Human Nutrition. Publ: Academic Press

Sasaki, S. (1996). Influence of sodium chloride on the levels of flavor compounds produced by shoyu yeast. Journal of Agricultural and Food Chemistry, **44** (10): 3273-3275

Scherz, H. and Senser, F. (1994). Souci, Fachmann and Kraut's Food Composition and Nutrition Tables. 5th Edition. Publ: CRC Press

Scientific Committee for Food (2001). Opinions on estragole and methyleugenol. http://europa.eu.int/comm/food/fs/sc/scf/outcome_en.html. 8th November.

Sharp, T.M. (1994). Development and significance of a novel food. Quorn mycoprotein. Voeding, **55** (1): 28-29

Sharp, T.M. (1999). The technical and economic barriers to the production of reduced fat bakery products. CCFRA R&D Report 86

Smith, A.K. and Circle, S.J. (1978). Soybeans: Chemistry & Technology. Volume 1 - Proteins. Publ: AVI

Smith, J. (1991). Food Additive User's Handbook. Publ: Blackie

Smith, D.B., Roddick, J.G. and Jones, J.L. (1996). Potato glycoalkaloids: some unanswered questions. Trends in Food Science and Technology, **7**: 126-131

Stryker, L. (1988). Biochemistry. 3rd Edition. Publ: WH Freeman

Taylor, S.L. and Hefle, S.L. (2001). Food allergies and other food sensitivities. Food Technology, **55** (9): 68-83

Thompson, T. (1997) Do oats belong in a gluten-free diet? Journal of the American Dietetic Association, **97**: 1413-1416.

Torres, R.L., Gonzalez, R.J., Sanchez, H.D., Osella, C.A., and de la Torre, M.A.G. (1999). Performance of rice varieties in making bread without gluten. Archivos Latinoamericanos de Nutricion, **49** (2): 162-165.

Troncone, R. and Auricchio, S. (1991). Gluten-densitive enteropathy (celiac disease). Food Reviews International, **7** (2): 205-231

University of Saskatchewan (1999). Production of legume pasta products by a high temperature extrusion process. United States Patent 5,989,620.

USDA Nutrient Data Laboratory. Food Composition Tables. www.nal.usda.gov/fnic/foodcomp/

Wagner, A.F. and Folkers, K. (1964). Vitamins and coenzymes. Publ: John Wiley and Sons

Wharton, B. and Booth, I. (2001). Fortification of flour with folic acid. British Medical Journal, **323**: 1198-1199

Whitney, E.N., Cataldo, C.B. and Rolfes, S.R. (1998). Understanding Normal and Clinical Nutrition. Wadsworth Publishing

Wilson, D. (2001). Marketing mycoprotein. The Quorn foods story. Food Technology, **55** (7): 48,50

Zimmet, P., Alberti, K.G.M.M. and Shaw, J. (2001). Global and societal implications of the diabetes epidemic. Nature, **414** (6865): 782-787

ABOUT CCFRA

The Campden & Chorleywood Food Research Association (CCFRA) is the largest membership-based food and drink research centre in the world. It provides wide-ranging scientific, technical and information services to companies right across the food production chain - from growers and producers, through processors and manufacturers to retailers and caterers. In addition to its 1500 members (drawn from over 50 different countries), CCFRA serves non-member companies, industrial consortia, UK government departments, levy boards and the European Union.

The services provided range from field trials of crop varieties and evaluation of raw materials through product and process development to consumer and market research. There is significant emphasis on food safety (e.g. through HACCP), hygiene and prevention of contamination, food analysis (chemical, microbiological and sensory), factory and laboratory auditing, training, publishing and information provision. To find out more, visit the CCFRA website at www.campden.co.uk

ABOUT ROYAL SOCIETY OF CHEMISTRY

The Royal Society of Chemistry is the UK Professional Body for chemical scientists and an international Learned Society for the chemical sciences with 46,000 members world-wide. It is a major international publisher of chemical information, supports the teaching of the chemical sciences at all levels and is a leader in bringing science to the public.

To find out more about the Royal Society of Chemistry visit the websites: www.rsc.org and www.chemsoc.org